North American
Range Plants

North American Range Plants

J. Stubbendieck
Stephan L. Hatch
and Kathie J. Hirsch

Third Edition

University of Nebraska Press
Lincoln
and London

Library of Congress Cataloging-in-Publication Data
Stubbendieck, James L.
    North American Range plants.
    Bibliography: p.
    Includes index.
    1. Range plants—North America—Identification.
2. Forage plants—North America—Identification.
3. Range plants—North America. 4. Forage plants—
North America. I. Hatch, Stephan L., 1945–
II. Hirsch, Kathie J., 1954–   III. Title.
SB193.3.N67S88   1986   582'.06   85-16491
ISBN   0-8032-9162-0 (pbk.)

Dedicated to range plant team coaches—
past, present, and future

Contents

Tribe	Scientific Name	

Forbs and Woody Plants

Scientific Name

Family	Scientific Name	

Acknowledgments

The authors wish to acknowledge the assistance of Susie Oldfather, Walter Schacht, Daniel Nosal, and T. Mark Hart for the collection and preparation of the information included in the First Edition. The following are acknowledged for their work in supplying additional information or reviewing the various sections of the Third Edition: Ms. Kelly Roberts and Dr. Kenneth Vogel, University of Nebraska–Lincoln; Dr. Clifford W. Morden, Mr. Chuck R. Coffey, and Mr. K. N. Gandhi, Texas A & M University; and Dr. Jesus Valdes-R., Universidad Autonoma Agraria Antonio Narro.

Dr. Jesus Valdes-R. is also recognized for furnishing the Mexican common names. Dr. Robert W. Knight, Texas A & M University, Mr. W. Robert Powell, University of California–Davis, and Dr. Mary Jane Davies, University of Alberta, suggested changes and additions to the Glossary.

Ms. Charlene Cunningham, University of Nebraska–Lincoln, and Ms. Catherine Garber, Texas A & M University, are acknowledged for typing the manuscript and proofing the Third Edition. Mrs. Catherine Mills, University of Nebraska–Lincoln, prepared the distribution maps.

The authors wish to acknowledge the Society for Range Management and its Executive Vice President, Mr. Peter V. Jackson, for permission to include the SRM International Range Plant Contest Rules. Canada Department of Agriculture and Texas A & M University Press are acknowledged for permission to reproduce specific drawings.

Finally, the authors wish to especially recognize Ms. Bellamy Parks, University of Nebraska–Lincoln, for preparation of additional drawings for the Third Edition. Her ability to combine her knowledge of plants and artistic skills will enhance the abilities of many to learn to identify these range plants.

Introduction

The need for a comprehensive reference to aid in the identification of the most important range plants of North America has long been recognized by ecologists, range managers, land managers, and other range professionals. In addition, university students and range plant identification teams have needed a single, primary resource for studying important range plant species. To this end, NORTH AMERICAN RANGE PLANTS was developed.

The 200 species in this book were selected because of their abundance, desirability, or noxious properties. This list of plants was developed over a 30-year period by coaches of range plant identification teams and faculty from the colleges and universities with "range management" programs. That formalized list is now The Master Plant List for The International Range Plant Identification Contest sponsored by the Society for Range Management.

Plant descriptions in this book include characteristics for their identification, a drawing of the plant or enlarged plant parts, and a general distribution map for North America. Each species description includes nomenclature; life span; origin; season of growth; inflorescence, flower or spikelet, vegetative, and growth characteristics. Forage value is estimated along with brief notes on habitat; livestock losses; and historic, food, and medicinal uses.

Grasses (POACEAE family) are described first, aligned by tribe, genus, and species in alphabetical order by rank. All other families follow in alphabetical order by rank for family, genus, and species with the exception of the ASTERACEAE family which follows the same format as the POACEAE family.

The grass (POACEAE) and composite (ASTERACEAE) families are treated by tribe to help the reader relate to smaller groups within these large, complex families. Recognition of plants within tribes builds a concept of tribal characteristics. When an unknown species of either family is encountered, knowledge of tribal alignments below family will aid in identification, thus reducing the time required for making an identification.

The classification system in F. W. Gould's *The Grasses of Texas,* Texas A & M University Press (1975), was followed for the grass tribal names. The tribal classification of the composites follows A. Cronquist's *Vascular Flora of the Southeastern United States,* University of North Carolina Press (1980).

Numerous authoritative floristic treatments from the rangeland areas of North America were consulted for species names and authorities. Selected synonyms, listing other names for the same species, are also given to help clarify the species concept used in this text (page 439). The synonyms will help in finding additional information from older floristic treatments, i.e., A. S. Hitchcock's *Manual of the Grasses of the United States* (revised by A. Chase), United States Department of Agriculture, Miscellaneous Publication 200 (1951).

Common names are given for the plants, but they may not include the common name used in a particular location. Spanish common names for most Mexican species are also listed.

The origin of each species is given as native or introduced. Origins of introduced taxa are given parenthetically. Many species are known to be introduced, while others are thought to have been. *Poa pratensis* L. is an example of a species listed as introduced, but that may be native to North America.

Season of growth is listed as cool, warm, or evergreen. Cool season plants complete most of their growth in the fall, winter, and spring, whereas warm season plants grow most when temperatures are the highest in the summer. The evergreen plants retain the ability to grow whenever climatic conditions are adequate.

Plant characteristics for each species are separated into categories to help in making comparisons between species. The intent is for these characteristics to be useful to the amateur botanist. In identifying plants, the conservative characteristics, those that are not greatly influenced by the environment, should be the basis for identification. These may include floral, spikelet, leaf, and inflorescence type, but may vary with the species. Pubescence, ligule lengths, and awn lengths are variable characters, and primary importance should not be placed on these when identifying grasses. Presence or absence of rhizomes is another variable characteristic used in grasses. This character is somewhat dependent upon moisture and other features of the habitat.

Forage values of the plants discussed in this book are relative values that vary with the animal utilizing the particular plant species. Values are determined on the basis of palatability, nutrient content, and the amount of forage produced by the plant species. These values may vary with the climatic conditions, when the forage is consumed, and the age class of each animal species utilizing the forage.

Losses from poisonous plants, one of the major problems facing the livestock industry, also are included in these plant descriptions. Losses on rangelands result in an annual loss in the millions of dollars, with the effects of poisonous plants varying from slightly reduced rate of gain to death of the animal. Losses which are easy to document, such as death, are not as economically important as the losses wherein growth rate or milk production is reduced. The brief discussions of livestock losses contained in this book gives the animal affected and the type of poison contained in the plant species.

In addition to the list of Selected Synonyms (page 439), this book includes a Glossary (page 423), list of Authorities (page 433), and a list of Selected References (page 443). This supplementary information will give the range student, professional range manager, and anyone else interested in plants a more complete knowledge of plants and a starting place in the literature to seek additional knowledge.

The information contained in NORTH AMERICAN RANGE PLANTS is by no means complete. The authors have settled for brevity with the expectation that this book is a starting point for those interested in range plant identification. Plant taxonomists and extension personnel in each locality can provide additional information on the plant species of interest.

1.

2.

3.

4.

5.

6.

7.

8.

9.

10.

11.

12.

13.

14.

15.

16.

17.

18.

19.

20.

21.

22.

23.

24.

Range Plants

Life span

Most range plants are classified as annuals or perennials. Annuals are plants that complete the life cycle in one growing season, while perennials generally live three or more years. Herbaceous perennials have aerial stems which die back each year while the underground parts remain alive. Perennial grasses and forbs are in this category. Woody perennials have aerial stems that remain alive throughout the year, although they may be dormant part of the time. Trees and shrubs are in this category. Biennial is a third life span category. Biennials require two growing seasons to complete the life cycle. Growth during the first year is only vegetative, and seed is produced during the second growing season. A relatively few plants fit into this category. All of the 200 plants in this book are classified as either an annual or a perennial, but a few of the perennials may occasionally act as biennials.

Origin

Most range plants are native, or plants that originated in North America. Introduced refers to plants which have been brought into North America from another continent and were adapted to conditions here. Several introduced species are valuable forage plants which were intentionally introduced for that purpose. Some introduced species were brought in for various reasons, but have escaped and are often troublesome weeds. Some plants were accidently introduced through contaminated crop seed, packing material, and ballast.

Classification

Botanical nomenclature refers to a system of naming plants. Plants are described and grouped according to their structure, particularly the flowering parts. The classification system involves a series of categories arranged to show the relationships and similarities of plants to one another. The classification system from general to specific is:

Division (Phylum)
Subdivision
Class
Order
Family
Tribe
Genus
Species

We will be concerned only with the last four parts of the classification system.

A. Family

A plant family is the basic division of plant orders. Morphological characteristics or similarities determine the family to which a plant belongs. Flowering characteristics are extremely important in the classification of family. All grasses have similar flowers and belong to the same family called POACEAE. Numbers of petals, sepals, stamens, and other flowering parts are the basic divisions. All families of vascular plants have the standard ending-ACEAE.

B. Tribe

A plant family may be divided into tribes. In this book the POACEAE and ASTERACEAE are the only families that are divided into tribes. An example is the ANDROPOGONEAE tribe of the POACEAE family.

C. Scientific name

There is only one correct scientific name for each plant. The scientific name consists of two main parts. The first part is the genus, and the second is the specific epithet. The authority is added for completeness and accuracy.

1. Genus

Classification of plants into genera is based on similarities in flowering and/or morphological characteristics, however, with more specific divisions. An example would be the genus *Schizachyrium* which is part of the ANDROPOGONEAE tribe of the POACEAE family. The first letter of the genus is capitalized, and the word is underlined or italicized.

2. Specific epithet

The second part of the scientific name is the specific epithet. A species is a group of similar plants, which is identified by the genus and specific epithet. This classification is based on differences in flowering and/or morphological characteristics. An example is the specific epithet *scoparium* of the species *Schizachyrium scoparium* which differs from all other species of *Schizachyrium* in morphological characteristics.

3. Authority
    The scientific name, for reasons of completeness and accuracy, is followed by the abbreviation or whole name of the person or persons who first applied that name to the plant. For example, (Michx.) Nash is the authority for *Schizachyrium scoparium*. The French botanist Andre Michaux (1746–1802) first described and applied the specific epithet to that specific plant, and the American agrostologist George Nash (1864–1921) later transferred the epithet to the genus *Schizachyrium*. A list of authorities begins on page 433.

4. Common name
    Common names have been given to most plants. Common names are usually simple and are often descriptive of the plant. Little bluestem is the common name of *Schizachyrium scoparium* (Michx.) Nash. Common names are only good in areas using the same language. In addition, one species may have several common names, and one common name may be applied to several species.

Only one correct scientific name exists for each plant. Nevertheless, the name for a given plant will change if that plant is reclassified or if it is discovered that an earlier name for it was validly published. Although date of publication is absolute, the assignment of rank and position in the classification process is a matter of taxonomic opinion, which is often annoying to the layman. An example of a scientific name change is that of little bluestem which now is *Schizachyrium scoparium* (Michx.) Nash and was formerly *Andropogon scoparius* Michx. This seems a recent change to most of us, but it was first proposed by Nash in 1903. The recent name change is a consensus of taxonomic opinion favoring Nash rather than Michaux. Names other than the correct one are synonyms. A list of selected synonyms begins on page 439.

A summary of the classification of little bluestem would be:

Family: POACEAE
  Tribe: ANDROPOGONEAE
    Genus: *Schizachyrium*
      Specific epithet: *scoparium*
Generally, it would be simply written:
  Schizachyrium scoparium
Or, more completely and correctly written:
  Schizachyrium scoparium (Michx.) Nash

Plant groups

Range plants may be divided into grasses, grasslike plants, forbs, and woody plants. These can be easily distinguished by certain characteristics.

Grasses have both hollow and solid stems with nodes. Leaves are two-ranked and have parallel veins, which are typical of monocots. Flowers are small and inconspicuous.

Grasslike plants resemble grasses, but generally have solid stems without elongated internodes. Leaf veins are parallel, but the leaves are three-ranked. Stems are often triangular, and the flowers are small and inconspicuous.

Forbs are herbaceous plants other than grasses and grasslike plants. They have solid stems, and generally have broad leaves that are usually net veined. Flowers are often large, colored, and showy.

Woody plants have secondary growth originating from aerial stems which live throughout the year, although they may be dormant part of the time. Leaves are often broad and net veined. Flowers are often showy. Both trees and shrubs fit into this category.

	Grasses	Grasslike sedges	Forbs	Shrubs
Stems	Jointed, Hollow or Pithy	Solid, not Jointed	Solid	Growth rings, Solid
Leaves	PARALLEL VEINS, STEM LEAF, LEAVES on 2 Sides	STEM LEAF, LEAVES on 3 Sides	"VEINS" are NETLIKE	
Flowers		Stamen, Ovary, MALE FEMALE (may be combined)	Usually Showy	
Example	Western Wheatgrass	Threadleaf Sedge	Yarrow	Big Sagebrush (twig)

Important Range Plant Groups

Morphology of grasses

The following is a series of drawings illustrating various morphological features of the grass plant. See the Glossary beginning on page 423 for definitions of terms.

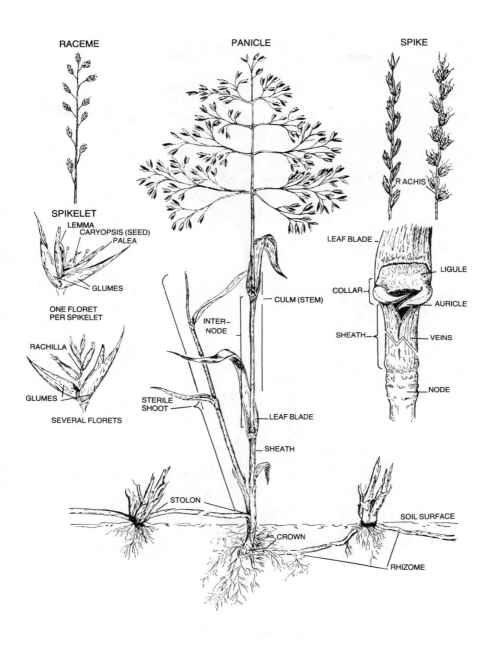

RACEME

PANICLE

SPIKE

SPIKELET

LEMMA
CARYOPSIS (SEED)
PALEA

GLUMES

ONE FLORET
PER SPIKELET

RACHILLA

GLUMES

SEVERAL FLORETS

RACHIS

LEAF BLADE

LIGULE

COLLAR

AURICLE

SHEATH

VEINS

NODE

CULM (STEM)

INTER-
NODE

STERILE
SHOOT

LEAF BLADE

SHEATH

STOLON

SOIL SURFACE

CROWN

RHIZOME

Grass Plant

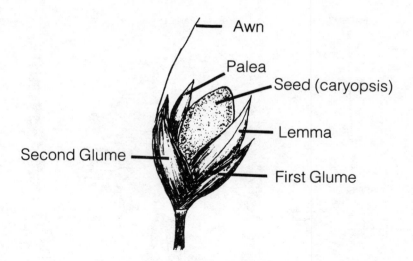

Awn

Palea

Seed (caryopsis)

Lemma

Second Glume

First Glume

Spikelet with only one floret

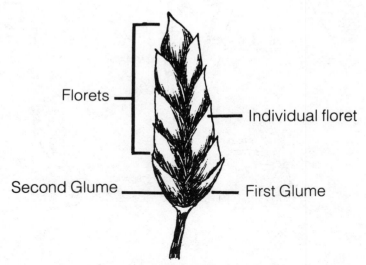

Florets

Individual floret

Second Glume

First Glume

Spikelet having several florets

Mature Spikelets

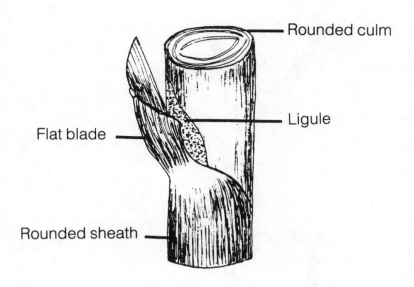

Rounded culm

Ligule

Flat blade

Rounded sheath

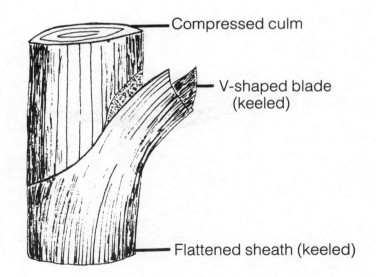

Compressed culm

V-shaped blade
(keeled)

Flattened sheath (keeled)

Grass Leaves and Culms

11

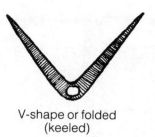

V-shape or folded
(keeled)

Involute

Flat

Cross Sections of Grass Leaves

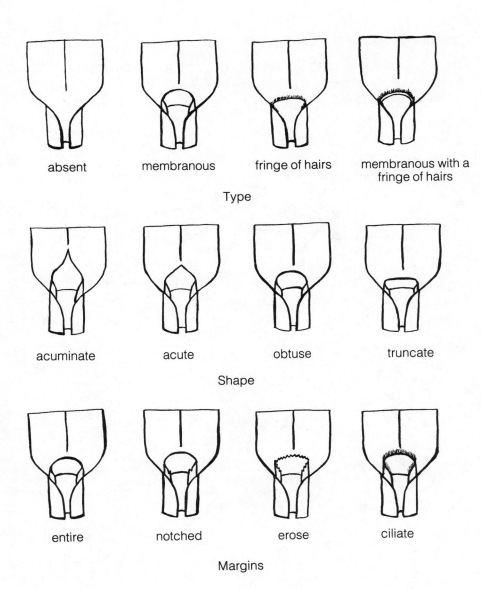

absent   membranous   fringe of hairs   membranous with a
                                        fringe of hairs

Type

acuminate   acute   obtuse   truncate

Shape

entire   notched   erose   ciliate

Margins

Type, Shape, and Margins of Grass Ligules

13

Morphology of forbs and woody plants

The following is a series of drawings illustrating the various morphological features of forbs and woody plants. See the Glossary beginning on page 423 for definitions of terms.

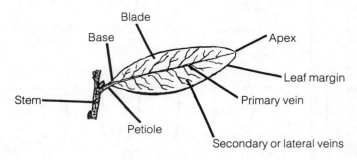

Parts of a Leaf

Simple leaf in
one piece

Palmately compound:
Spreading from the tip of
the stem like fingers
from the palm of a hand

Pinnately compound:
Leaflets arranged on both
sides of the petiole

Opposite:
Two leaves to a node

Whorl:
A circle of leaves
at the same
joint or node

Alternate:
One leaf to a node

Leaf Arrangement

simple

pinnately
compound

bipinnately
compound

awl-shaped
needles

scale-like
needles

pine
needles

Leaf Types

Linear:
Long and narrow with parallel sides.

Lanceolate:
Much longer than wide and tapering upwards from the middle.

Oblong:
Longer than broad with parallel sides.

Elliptical:
Broadest in the middle, equally rounded at the ends.

Orbicular:
Round.

Ovate:
Egg-shaped, broadest near the base.

Obovate:
Egg-shaped with broadest end at the top.

Cordate:
Heart-shaped.

Deltoid:
Triangular.

Sagittate.
Two basal lobes directed backward.

Cuneate.
Tapering to point of attachment.

Leaf Shape

| Entire | Dentate | Toothed or Serrate | Sinuate or Wavy | Doubly Serrate | Lobed | Incised |

Leaf Margins

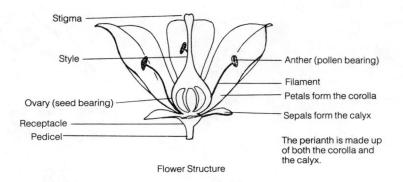

Stigma

Style

Anther (pollen bearing)

Filament

Petals form the corolla

Ovary (seed bearing)

Sepals form the calyx

Receptacle

Pedicel

The perianth is made up
of both the corolla and
the calyx.

Flower Structure

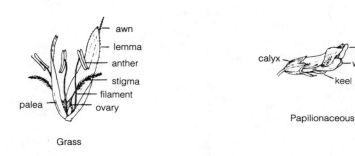

awn

lemma

anther

stigma

filament

palea

ovary

Grass

calyx

banner (standard)

wing

keel

Papilionaceous

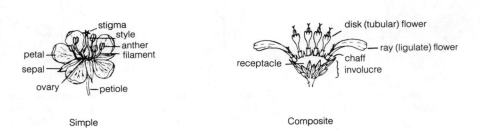

stigma

style

anther

filament

petal

sepal

ovary

petiole

Simple

disk (tubular) flower

ray (ligulate) flower

receptacle

chaff

involucre

Composite

Kinds of Flowers

Separate petals

Fused corolla

Spurred:
A spur is a hollow,
tubular projection
on a flower

Regular flower
radially symmetric

Irregular flower
bilaterally symmetric

Spike:
An unbranched inflorescence
with flowers sessile
on the rachis

Catkin:
The scaly spike or
raceme of a woody plant

Panicle:
An inflorescence with
a main axis and
subdivided branches

Raceme:
An inflorescence with
flowers pediceled
on the rachis

Terminal:
Single blossom at
the top of a
stem or scape

Umbel:
A flower cluster in which
all flowers arise from a
common terminal point

Compound umbel

Types of Flowers and Inflorescences

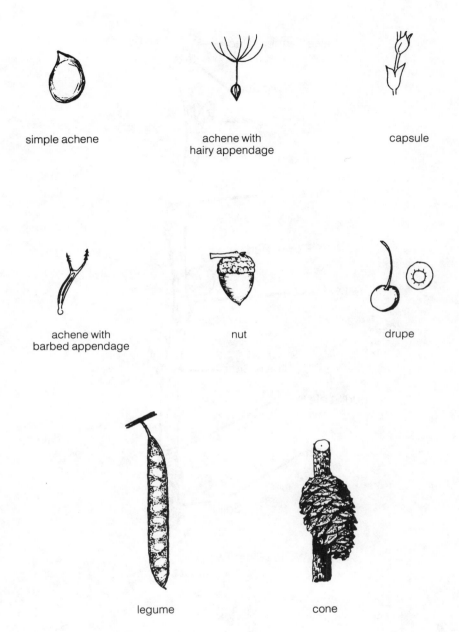

simple achene

achene with
hairy appendage

capsule

achene with
barbed appendage

nut

drupe

legume

cone

Types of Fruits and Seeds

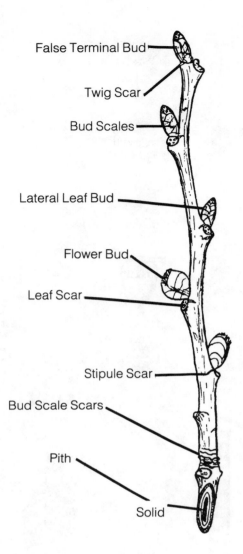

False Terminal Bud

Twig Scar

Bud Scales

Lateral Leaf Bud

Flower Bud

Leaf Scar

Stipule Scar

Bud Scale Scars

Pith

Solid

Woody Plant Twig

Alternate

Whorled

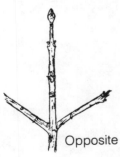

Opposite

Types of Twig Branches

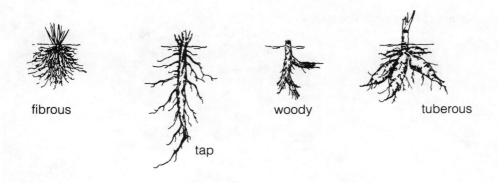

fibrous

woody

tuberous

tap

stolon

rhizome

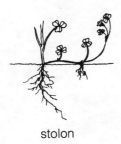

caulescent

acaulescent

Types of Roots and Stems

Saltgrass (*Distichlis spicata*)

Notes:

Tribe	AELUROPODEAE
Species	*Distichlis spicata* (L.) Greene
Common Name	Saltgrass (zacate salado)
Life Span	Perennial
Origin	Native
Season	Warm

INFLORESCENCE CHARACTERISTICS

type   panicle (1–6 cm long), contracted

spikelets   5- to 9-flowered (0.6–1.0 cm long); lemma acute (3–6 mm long), edge yellow and coarse; palea soft, narrowly winged

glumes   unequal, acute, glabrous, 3- to 9-nerved

other   dioecious; staminate spikelets straw-colored, pistillate spikelets green

VEGETATIVE CHARACTERISTICS

culm   decumbent to erect (10–40 cm tall), internodes short and numerous

sheath   closely overlapping

blade   conspicuously 2-ranked (distichous), flat to involute (generally less than 10 cm long), sharp-pointed, tightly curled

ligule   membranous (0.5 mm long), ciliate, truncate

other   extensively creeping and scaly rhizomes

GROWTH CHARACTERISTICS   starts growth in early summer, slow growth rate, remains green until fall, few seeds produced, reproduction mostly from rhizomes

FORAGE VALUE   little for livestock or wildlife, seldom grazed if other grasses are available

HABITAT   seashores and alkaline inland soils

Big bluestem (*Andropogon gerardii*)

Tribe	ANDROPOGONEAE
Species	*Andropogon gerardii* Vitman
Common Name	Big bluestem (popotillo gigante)
Life Span	Perennial
Origin	Native
Season	Warm

INFLORESCENCE CHARACTERISTICS

type   panicle of 2–6 digitate racemes (5–10 cm long), commonly 3, sessile on a terminal peduncle

spikelets   paired, lower spikelet sessile and perfect (0.7–1.0 cm long), paired spikelets nearly equal length, pedicelled spikelet sterile

awns   lemma of sessile spikelet awned (1–2 cm long), geniculate and tightly twisted below

glumes   first glume slightly grooved or dished

other   often purplish, sometimes yellowish

VEGETATIVE CHARACTERISTICS

culm   erect (1–2 m tall), robust, glabrous, sparingly branched toward summit

sheath   compressed, purplish at base, lower sheaths sometimes with villous long and soft hairs, hyaline margins

blade   flat (10–50 cm long and 0.5–1.0 cm wide), lower blades often villous, margins scabrous

ligule   short ciliate (1.0–2.5 mm long), square collar

other   sometimes with short rhizomes

GROWTH CHARACTERISTICS   growth cycle is 3–4 months, numerous leaves produced in late spring, growing points stay near ground level until late summer

FORAGE VALUE   excellent and highly palatable to all classes of livestock but it becomes coarse late in the season

HABITAT   prairies, dry soils, open woods, and wet overflow sites

Broomsedge bluestem (*Andropogon virginicus*)

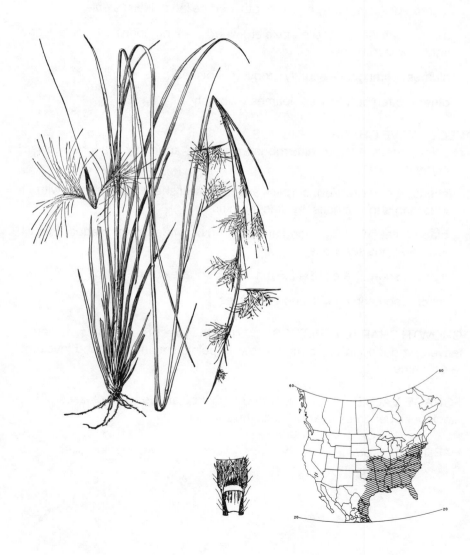

Tribe	ANDROPOGONEAE
Species	*Andropogon virginicus* L.
Common Name	Broomsedge bluestem (popotillo)
Life Span	Perennial
Origin	Native
Season	Warm

INFLORESCENCE CHARACTERISTICS

type   panicle of 2−4 racemose branches (2−3 cm long)

spikelets   paired, lower spikelet sessile and perfect (3−4 mm long)

awns   lemma with delicate straight awn (1−2 cm long)

glumes   yellow to green (3−4 mm long)

other   base of branches enclosed in an inflated, tawny spathe

VEGETATIVE CHARACTERISTICS

culm   small tufts (0.5−1.0 m tall), flat basal nodes, glabrous, branched above

sheath   overlapping, lower sheaths compressed, keeled, margins hairy

blade   flat or folded (2−5 mm wide), pilose on upper surface, orange to straw-colored at maturity

ligule   membranous (1.0 mm long), ciliate, truncate

GROWTH CHARACTERISTICS   starts growth when daytime temperatures average 16−17° C, grows in infertile soils but is not shade tolerant

FORAGE VALUE   low, except in early growth stages

HABITAT   open ground, old fields, open woods, sterile hills and sandy soil

Silver bluestem (*Bothriochloa saccharoides*)

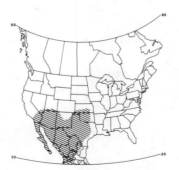

28

Tribe	ANDROPOGONEAE
Species	*Bothriochloa saccharoides* (Sw.) Rydb.
Common Name	Silver bluestem (silver beardgrass, popotillo plateado)
Life Span	Perennial
Origin	Native
Season	Warm

INFLORESCENCE CHARACTERISTICS

type   panicle of 6 or more racemose branches, inflorescence long-exserted (7–15 cm long), elongate, terminal

spikelets   paired, lower spikelet sessile and perfect (3 mm long)

awns   lemma with delicate awn (0.8–1.8 cm long)

glumes   firm, papery, first 2-keeled, second 1-keeled and 3-nerved

other   silvery-white in color, rachis joints and pedicels long-villous, pedicel is grooved, pedicel cross section is dumbbell-shaped

VEGETATIVE CHARACTERISTICS

culm   often crooked (0.6–1.3 m tall), nodes without tuft of long white hair

sheath   oval, glabrous, keeled near collar, collar with long hairs on margin

blade   flat (3–9 mm wide), linear (8–20 cm long), tapering to a point, glaucous, broad midrib, white margins

ligule   membranous (1–3 mm long), toothed

GROWTH CHARACTERISTICS   starts growth in spring when daytime temperatures reach 21–24° C, inflorescence emerges 3–4 weeks later

FORAGE VALUE   fair for all classes of livestock and wildlife

HABITAT   prairies and rocky slopes

Tanglehead (*Heteropogon contortus*)

Tribe	ANDROPOGONEAE
Species	*Heteropogon contortus* (L.) Beauv. *ex* R. & S.
Common Name	Tanglehead (zacate colorado)
Life Span	Perennial
Origin	Native
Season	Warm

INFLORESCENCE CHARACTERISTICS

type   spicate raceme (3–8 cm long), solitary, 1-sided

spikelets   paired, imbricate; both spikelets at base of inflorescence are staminate or neuter, sessile spikelet perfect

awns   lemma awn prominent (5–12 cm long), dark, bent, tangled

glumes   brownish-hispid, glumes of staminate spikelets hirsute or hispid

VEGETATIVE CHARACTERISTICS

culm   tufted (20–80 cm tall), flat, branched above

sheath   flat to compressed keeled, flat base, margin glandular and hairy

blade   glaucous (6–20 cm long and 4–6 mm wide), adaxial midvein prominent, margin white glandular, tips and base red at maturity

ligule   membranous (1 mm long), ciliate

GROWTH CHARACTERISTICS   starts growth in early spring and matures in August

LIVESTOCK LOSSES   awns may be troublesome, especially to sheep

FORAGE VALUE   fair to good for cattle and horses before maturity; due to coarseness, it is of little value to sheep

HABITAT   rocky hills and canyons

Little bluestem (*Schizachrium scoparium*)

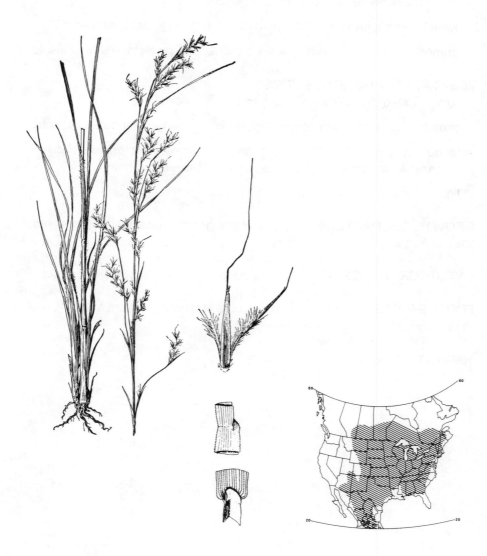

32

Tribe	ANDROPOGONEAE
Species	*Schizachyrium scoparium* (Michx.) Nash
Common Name	Little bluestem (prairie beardgrass, zacate colorado)
Life Span	Perennial
Origin	Native
Season	Warm

INFLORESCENCE CHARACTERISTICS

type   racemes (2.5–5.0 cm long), solitary

spikelets   paired (6–8 mm long), sessile spikelet perfect, pedicelled spikelet sterile, rachis and pedicels pilose

awns   lemma of sessile spikelet awned (1.2 cm long), awn bent and twisted

glumes   thickened, rounded on the back

other   peduncle included in sheath, inflorescence jointed, breaking apart at the joints

VEGETATIVE CHARACTERISTICS

culm   decumbent (0.5–1.5 m tall), tufted, flat, leafy base, green, glaucous

sheath   flattened, laterally keeled, glabrous

blade   linear (8–25 cm long and 2–6 mm wide), acuminate, glabrous to hispid, scabrous on upper surface and margins

ligule   membranous (1–3 mm long), truncate, ciliate

other   short rhizomes

GROWTH CHARACTERISTICS   starts growth in late spring, matures in early fall, seed matures October–November

FORAGE VALUE   fair to good while young and tender; after heads mature, forage is fair for cattle and horses but is too coarse for sheep or goats

HABITAT   prairies, open woods; dry hills and fields

Indiangrass (*Sorghastrum nutans*)

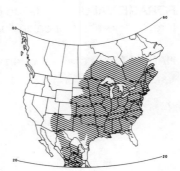

Tribe	ANDROPOGONEAE
Species	*Sorghastrum nutans* (L.) Nash
Common Name	Indiangrass (yellow indiangrass, zacate indio)
Life Span	Perennial
Origin	Native
Season	Warm

INFLORESCENCE CHARACTERISTICS

type   panicle (15–30 cm long), dense

spikelets   paired (6–8 mm long), pedicelled spikelet absent

awns   lemma of perfect spikelet awned (1.2–1.8 cm long), geniculate

glumes   leathery, brown or yellow, first hirsute with edges inflexed over the second

other   inflorescence yellow or gold; grayish-hirsute summits of branchlets, rachis joints, and pedicels

VEGETATIVE CHARACTERISTICS

culm   erect (1–2 m tall), nodes pubescent

sheath   round or sometimes flattened, pilose, hairs long near collar

blade   flat or somewhat keeled (10–30 cm long and 0.5–1.0 cm wide), constricted at the base, midvein conspicuous dorsally

ligule   membranous (2–5 mm long), deeply notched (horn-like or leaf-like)

other   short, scaly rhizomes

GROWTH CHARACTERISTICS   starts growth in midspring from short rhizomes, matures from September to November, also reproduces by seed

FORAGE VALUE   palatable for cattle and horses throughout the summer, but it does not cure well and is generally considered only moderately palatable after maturity and fair forage for winter grazing

HABITAT   prairies, bottomlands, open woods, and meadows

Eastern gamagrass (*Tripsacum dactyloides*)

Notes:

Tribe	ANDROPOGONEAE
Species	*Tripsacum dactyloides* (L.) L.
Common Name	Eastern gamagrass (zacate maicero)
Life Span	Perennial
Origin	Native
Season	Warm

INFLORESCENCE CHARACTERISTICS

type   panicle of 1–3 racemes (12–25 cm long), terminal

spikelets   unisexual, staminate above (0.7–1.1 cm long), pistillate below (0.7–1.0 cm long), bead-like, imbedded in rachis

glumes   generally firm and have the texture of stiff writing paper

other   pistillate portion ¼ or less of the entire length

VEGETATIVE CHARACTERISTICS

culm   tufted (1.5–3.0 m tall)

sheath   rounded, smooth, glabrous

blade   flat (30–75 cm long and 1–3 cm wide), scabrous on margins and above

ligule   membranous (1.0–2.5 mm long), ciliate, truncate

other   thick, knotty rhizomes

GROWTH CHARACTERISTICS   few seeds are produced, most reproduction from rhizomes, most growth is in the spring and it remains green until after frost

FORAGE VALUE   excellent forage for all classes of livestock throughout the year

HABITAT   swales, banks of streams, and moist places

Red threeawn (*Aristida longiseta*)

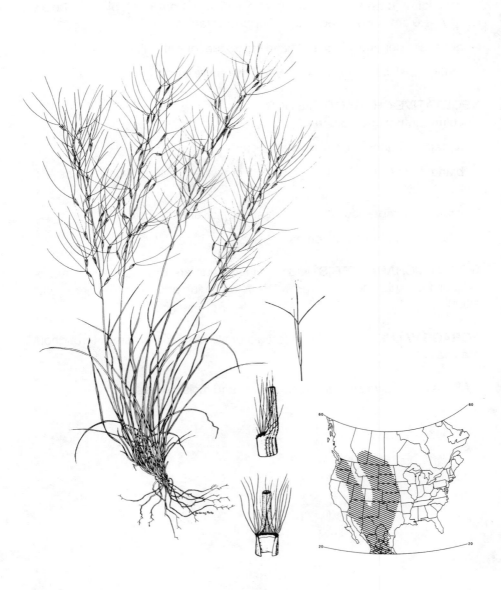

Tribe	ARISTIDEAE
Species	*Aristida longiseta* Steud.
Common Name	Red threeawn (purple threeawn, dogtown grass, tres barbas)
Life Span	Perennial
Origin	Native
Season	Warm

INFLORESCENCE CHARACTERISTICS

type   panicle (2–10 cm long), erect, somewhat flexuous, purple to red

spikelets   1-flowered

awns   lemma awn column branches into 3 awns (6–8 cm long), equal, divergent

glumes   first glume (0.8–1.0 cm long), ½ as long as second glume

VEGETATIVE CHARACTERISTICS

culm   tufted (10–35 cm tall), glabrous

sheath   glabrous to weakly scabrous, collar pubescent with a tuft of long and soft hairs on either side

blade   involute (2–12 cm long and 1–2 mm wide), sharp pointed, curved, scabrous, mostly basal

ligule   membranous (0.5–1.0 mm long), ciliate

GROWTH CHARACTERISTICS   starts growth in late spring, strong competitor, grows best after wet summer and fall followed by a dry winter

LIVESTOCK LOSSES   awns may decrease fleece value; awns may also injure the mouths and nostrils of grazing animals

FORAGE VALUE   grazed only in early growth stages before awn development

HABITAT   dry sandy soils, hillsides, particularly on disturbed sites

Prairie threeawn (*Aristida oligantha*)

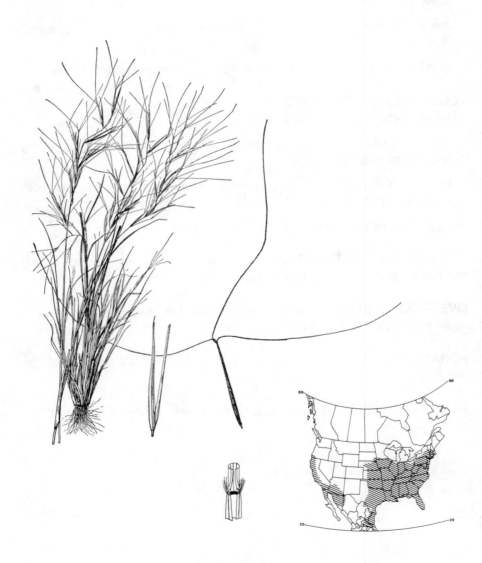

40

Tribe	ARISTIDEAE
Species	*Aristida oligantha* Michx.
Common Name	Prairie threeawn (oldfield threeawn, tres barbas)
Life Span	Annual
Origin	Native
Season	Warm

INFLORESCENCE CHARACTERISTICS

type   panicle (10–20 cm long), loose

spikelets   1-flowered, short pedicelled

awns   lemma awn column branches into 3 awns (4–7 cm long), nearly equal, divergent

glumes   nearly equal (2–3 cm long), tapering to awn-like point

VEGETATIVE CHARACTERISTICS

culm   decumbent (15–80 cm long), branches at lower nodes

sheath   rounded on back, glabrous or a few hairs

blade   few (10–20 cm long and 2–4 mm wide), flat or loosely involute

ligule   membranous, minute, fringed

GROWTH CHARACTERISTICS   growth starts in early spring, completes life cycle in 2 months

LIVESTOCK LOSSES   long awns can cause injury to livestock

FORAGE VALUE   brief period of fair to poor forage in early growth stages, otherwise worthless

HABITAT   open, dry ground on disturbed areas

Spikebent (*Agrostis exarata*)

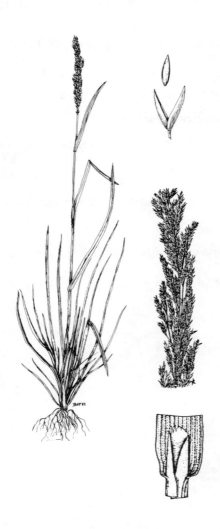

Tribe	AVENEAE
Species	*Agrostis exarata* Trin.
Common Name	Spikebent (spike redtop)
Life Span	Perennial
Origin	Native
Season	Cool

INFLORESCENCE CHARACTERISTICS

type   panicle (5−30 cm long), narrow, contracted

spikelets   1-flowered; lemmas ⅔ as long as glumes

awns   lemma may have a small delicate awn from the middle of the back

glumes   nearly equal (2−4 mm long)

other   panicle branches densely flowered at base, branches commonly in dense whorls

VEGETATIVE CHARACTERISTICS

culm   tufted, erect (0.2−1.2 m tall), somewhat bent at base, smooth

sheath   smooth to somewhat scabrous

blade   flat (4−20 cm long and 0.2−1.0 cm wide), usually scabrous

ligule   membranous (6 mm long), erose, acuminate, decurrent

other   leaves mostly basal

GROWTH CHARACTERISTICS   growth starts in spring and flowers in July or August, reproduces by seeds

FORAGE VALUE   excellent for all classes of livestock as well as for elk and deer, grazed throughout the summer

HABITAT   moist open ground at low and medium altitudes

Redtop bent (*Agrostis stolonifera*)

Tribe	AVENEAE
Species	*Agrostis stolonifera* L.
Common Name	Redtop bent (creeping bentgrass, redtop)
Life Span	Perennial
Origin	Introduced (from Europe)
Season	Cool

INFLORESCENCE CHARACTERISTICS

type panicle (5–30 cm long), pyramidal, spreading

spikelets 1-flowered; lemma blunt-tipped

glumes nearly equal, longer than lemma, pointed

other lower panicle branches whorled, not densely flowering at base, flowering nodes red

VEGETATIVE CHARACTERISTICS

culm erect (1.0–1.5 m tall) or sometimes decumbent

sheath round, glabrous, frequently purple to red

blade flat (4–20 cm long and 0.5–1.0 cm wide) at base, narrow pointed midvein prominent abaxially, margins barbed

ligule membranous (3–5 mm long), acute, erose

other creeping rhizomes

GROWTH CHARACTERISTICS   starts growth in early spring, matures by August, reproduces by rhizomes and seeds

FORAGE VALUE   good to very good for cattle and horses, and fairly good to good for sheep

HABITAT   seeded in pastures and moist meadows, widely naturalized

Slender oat (*Avena barbata*)

Notes:

Tribe	AVENEAE
Species	*Avena barbata* Pott. *ex* Link
Common Name	Slender oat (avena)
Life Span	Annual
Origin	Introduced (from Europe)
Season	Cool

INFLORESCENCE CHARACTERISTICS

type panicle (20–40 cm long), open, loose

spikelets 2-flowered on curved capillary pedicels, drooping; lemmas with stiff, red hairs to the middle, tapering to 2 long and narrow teeth (3–4 mm long)

awns lemmas awned from back (3–4 cm long), twisted, geniculate

glumes coarse-ribbed, 7-nerved (2.0–2.5 cm long)

other pedicels curved, smaller than *Avena fatua*

VEGETATIVE CHARACTERISTICS

culm erect (0.3–1.2 m tall)

sheath rounded to somewhat keeled, glabrous or nearly so

blade flat to slightly keeled (10–40 cm long and 0.5–1.0 cm wide), scabrous on both sides, margin often pilose

ligule membranous (4 mm long), acute to acuminate, erose margin

GROWTH CHARACTERISTICS   germinates in late fall or early winter, most growth is in early spring, flowers March–June

FORAGE VALUE   good during the winter and spring growth period, low palatability after it matures

HABITAT   foothill ranges, fields, and waste places

Wild oat (*Avena fatua*)

48

Tribe	AVENEAE
Species	*Avena fatua* L.
Common Name	Wild oat (avena)
Life Span	Annual
Origin	Introduced (from Europe)
Season	Cool

INFLORESCENCE CHARACTERISTICS

type   panicle (10–30 cm long), open

spikelets   2- or 3-flowered or rarely 4-flowered; lemma with long hairs and minutely 2-toothed at tip

awns   lemma awned from back (3–4 cm long), reddish-brown to black; twisted, geniculate

glumes   papery, coarse-ribbed, 9-nerved (1.5–2.5 cm long), pointed, longer than florets

other   panicle branches unequal, horizontally spreading to ascending

VEGETATIVE CHARACTERISTICS

culm   erect (0.3–1.3 m tall), small tufts, stout, smooth

sheath   rounded to somewhat keeled, margins broad and thick, glabrous to pubescent, collar pilose on front margin

blade   flat (10–30 cm long and 0.5–1.0 cm wide), scabrous on both sides, margins pilose especially near base

ligule   membranous (4 mm long), acuminate to rounded, erose margin

GROWTH CHARACTERISTICS   germinates in early winter, most growth in early spring, flowers March–May, reproduces by seeds

FORAGE VALUE   good to excellent for all classes of livestock until after the seed is mature

HABITAT   valley and foothill ranges, cultivated soil and waste places

Pine reedgrass (*Calamagrostis rubescens*)

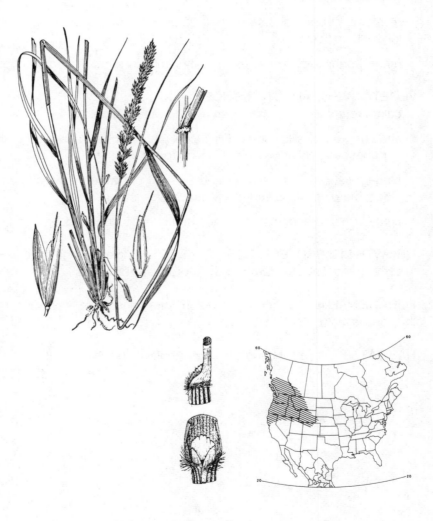

Tribe	AVENEAE
Species	*Calamagrostis rubescens* Buckl.
Common Name	Pine reedgrass (pinegrass)
Life Span	Perennial
Origin	Native
Season	Cool

INFLORESCENCE CHARACTERISTICS

type   panicle (7–15 cm long), dense, spike-like, interrupted

spikelets   1-flowered; lemma pale, nearly as long as the glumes, lemma with tuft of hair at base

awns   lemma awned from back near base, delicate, exserted at side between the glumes, shorter than glumes, once geniculate

glumes   narrow (4–5 mm long), nearly equal, acuminate

other   inflorescence sometimes red to purple

VEGETATIVE CHARACTERISTICS

culm   tufted (0.6–1.0 m tall), slender, nodes may be dark, and stems may be red

sheath   smooth, distinctly veined, often purple at base, pubescent collar

blade   flat or enrolled at tip (7–30 cm long and 2–4 mm wide), ascending with curved or drooping tips, leaves mostly basal

ligule   membranous (3 mm long), acute, erose

other   extensively creeping rhizomes, leaves mainly basal

GROWTH CHARACTERISTICS   starts growth in early spring, flowers in July or August, reproduces by seed

FORAGE VALUE   poor to fair for sheep, fair for cattle and horses, fair for elk

HABITAT   open pine woods, prairies, and stream banks

Tufted hairgrass (*Deschampsia caespitosa*)

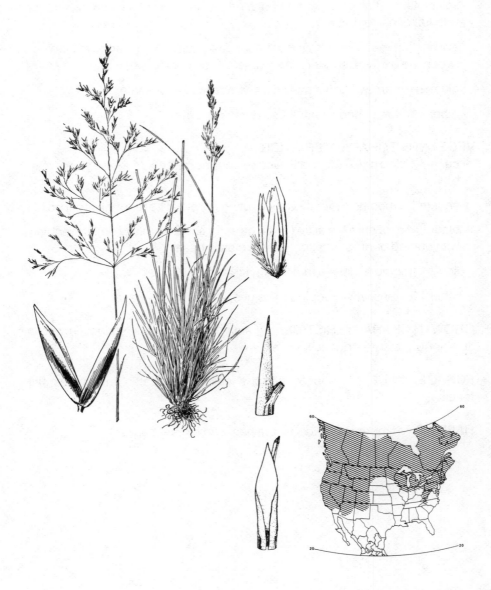

Tribe	AVENEAE
Species	*Deschampsia caespitosa* (L.) Beauv.
Common Name	Tufted hairgrass (saltandpepper grass)
Life Span	Perennial
Origin	Native
Season	Cool

INFLORESCENCE CHARACTERISTICS

type   panicle (10-25 cm long), open, shiny black when mature, panicle branches hair-like

spikelets   2-flowered (4−7 mm long); lemma thin with ragged tip

awns   lemma awned from back near base, length varies from short to twice as long as spikelet, weakly geniculate

glumes   slightly longer than lemmas, pointed

other   spikelets may have a bicolored appearance and be dark at maturity, hence the name saltandpepper grass

VEGETATIVE CHARACTERISTICS

culm   erect (0.6−1.2 m tall), dense tufts

sheath   smooth, keeled, veins prominent

blade   flat or folded (5−30 cm long and 1.5−4.0 mm wide), contracted at collar, ridged and scabrous dorsally, margins scabrous

ligule   membranous (0.4−1.0 cm long), acuminate

other   leaves mostly basal, departing at 45° angles

GROWTH CHARACTERISTICS   reproduces by seed, flowers from July to September

FORAGE VALUE   good to excellent for all classes of livestock and fair to good for wildlife

HABITAT   bogs and wet sites as well as in the spruce-fir belt and above timberline

Spike oat (*Helictotrichon hookeri*)

Tribe	AVENEAE
Species	*Helictotrichon hookeri* (Scribn.) Henr.
Common Name	Spike oat
Life Span	Perennial
Origin	Native
Season	Cool

INFLORESCENCE CHARACTERISTICS

type   panicle, long-exserted, narrow (5−10 cm long), branches erect or ascending, lower branches with 2 spikelets, rachilla villous

spikelets   3- to 6-flowered (1.5 cm long); lemmas firm (1.2 cm long), toothed at the apex, rounded on the back; palea well developed; callus short bearded

awns   awned from the middle of the back of the lemma; geniculate, twisted (1.0−1.5 cm long)

glumes   equal or the first shorter (1.2−1.5 cm long), acute apex, thin and membranous

VEGETATIVE CHARACTERISTICS

culm   densely tufted (50−80 cm tall)

sheath   glabrous, strongly keeled

blade   firm, flat, or folded (1−4 mm wide); margin somewhat thickened, greenish-white

ligule   membranous, erose-lacerate (1−3 mm long)

GROWTH CHARACTERISTICS   starts growth in early spring, flowers June−July, seed matures July−August, reproduces by seed and tillers

FORAGE VALUE   good forage for cattle, fair for sheep and deer

HABITAT   dry slopes and prairies

Prairie junegrass (*Koeleria pyramidata*)

56

Tribe	AVENEAE
Species	*Koeleria pyramidata* (Lam.) Beauv.
Common Name	Prairie junegrass (junegrass, Koelersgrass)
Life Span	Perennial
Origin	Native
Season	Cool

INFLORESCENCE CHARACTERISTICS

type   panicle (3–18 cm long and 1–3 cm wide), usually contracted

spikelets   2- to 5-flowered; lemma narrow (6 mm long), lanceolate, sharp-pointed, scabrous

glumes   about equal, unlike in shape; first 1-nerved, narrow; second 3-nerved and broadened above the middle, shiny and translucent, shorter than first floret

other   inflorescence may taper at both ends, bottom panicle branches may be pubescent

VEGETATIVE CHARACTERISTICS

culm   erect (20–60 cm tall), tufted, few fine hairs just below the inflorescence

sheath   retrorsely pubescent, distinctly veined, collar pilose on margin

blade   flat or involute (3–25 cm long and 1–3 mm wide), glabrous or pubescent on back, may have long hairs near collar, blunt, uniformly ribbed

ligule   membranous (0.5–1.0 mm long), truncate, ciliate

other   leaves mostly basal

GROWTH CHARACTERISTICS   starts growth in early spring, flowers in June and July, produces seed through September, high volume of seed with low viability

FORAGE VALUE   excellent for all classes of livestock, although its forage production is low, and good for wildlife in spring and in the fall after curing

HABITAT   prairie, open woods, and sandy soil

Hardinggrass (*Phalaris tuberosa*)

58

Tribe	AVENEAE
Species	*Phalaris tuberosa* L.
Common Name	Hardinggrass
Life Span	Perennial
Origin	Introduced (through Australia, native to Mediterranean area)
Season	Cool

INFLORESCENCE CHARACTERISTICS

type   panicle (5–15 cm long), contracted, dense, spike-like

spikelets   flat, 1 fertile floret (4 mm long), 2 sterile florets below fertile floret

glumes   broad (5–6 mm long), narrowly winged on upper ⅔

VEGETATIVE CHARACTERISTICS

culm   erect (0.7–1.5 m tall), stout, swollen or bulb-like base

sheath   glabrous

blade   glabrous (25–40 cm long and 0.6–1.0 cm wide)

ligule   membranous (6–8 mm long), obtuse

other   strong, loosely branching rhizomes

GROWTH CHARACTERISTICS   starts growth in late fall, flowers by mid-June, reproduces by seeds and rhizomes

FORAGE VALUE   excellent for all classes of livestock and big game in early growth stages, tends to become rank and tough with maturity

HABITAT   seeded on grainland, cleared foothill ranges, and cleared brushland for pasture and hay production

Alpine timothy (*Phleum alpinum*)

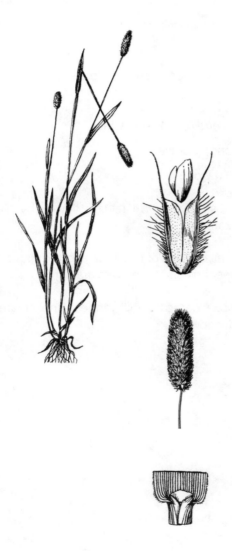

Tribe	AVENEAE
Species	*Phleum alpinum* L.
Common Name	Alpine timothy
Life Span	Perennial
Origin	Native
Season	Cool

INFLORESCENCE CHARACTERISTICS

type   panicle (2−5 cm long), dense, spike-like, ellipsoid, cylindrical

spikelets   1-flowered, small, somewhat flattened; lemma shorter than glumes

awns   glumes awned from tips (2 mm long)

glumes   hispid-ciliate on keel (5 mm long)

other   the inflorescence is usually from 1½ to 3 times as long as wide, often purplish in color

VEGETATIVE CHARACTERISTICS

culm   erect or decumbent (15−60 cm tall), internodes exposed

sheath   upper sheaths inflated, prominently veined

blade   flat (2−15 cm long and 4−8 mm wide), tapering, margins and dorsal surface scabrous

ligule   membranous (2−4 mm long), truncate or obtuse

other   base of culm may be decumbent and creeping

GROWTH CHARACTERISTICS   starts growth in early spring, reproduces by seeds and tillers

FORAGE VALUE   excellent for all classes of livestock and big game

HABITAT   mountain meadows and wet places

Timothy (*Phleum pratense*)

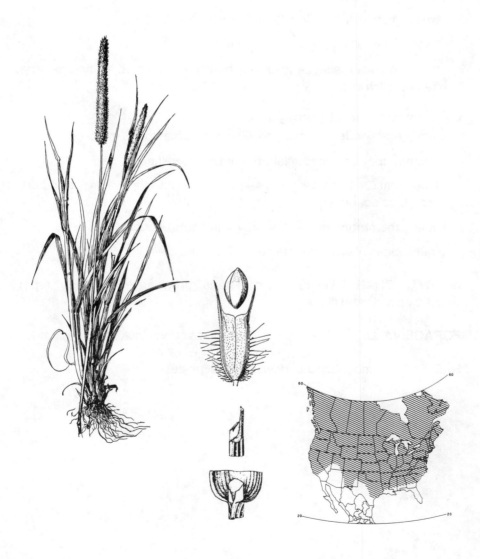

62

Tribe	AVENEAE
Species	*Phleum pratense* L.
Common Name	Timothy
Life Span	Perennial
Origin	Introduced (from Europe)
Season	Cool

INFLORESCENCE CHARACTERISTICS

type   panicle (3–20 cm long), contracted, dense, spike-like

spikelets   1-flowered, flattened, small; lemma about ½ as long as the glumes

awns   glumes awned from tips (1 mm), less than ½ the length of the glumes

glumes   equal (2–3 mm long), pectinate-ciliate on keel

other   inflorescence is several times longer than wide

VEGETATIVE CHARACTERISTICS

culm   erect (0.5–1.2 m tall), tufted, forming large clumps

sheath   round, clasping and closed around the nodes, distinctly veined, often purple at the base

blade   flat (5–30 cm long and 5–8 mm wide), tapering to a sharp point, distinctly veined, retrorsely scabrous margins

ligule   membranous (2–3 mm long), obtuse or acute

other   swollen or bulb-like base

GROWTH CHARACTERISTICS   reproduces from seed, cold tolerant, poor drought tolerance, responds to nitrogen fertilizer

FORAGE VALUE   produces leafy, palatable hay for cattle and horses and is seeded as a pasture grass; not tolerant of heavy grazing

HABITAT   seeded in pastures and meadows; commonly escaped from cultivation into roadsides, fields, and waste places

Spike trisetum (*Trisetum spicatum*)

64

Tribe	AVENEAE
Species	*Trisetum spicatum* (L.) Richt.
Common Name	Spike trisetum
Life Span	Perennial
Origin	Native
Season	Cool

INFLORESCENCE CHARACTERISTICS

type   panicle (2–15 cm long), dense, spike-like

spikelets   3-flowered (4–6 mm long); lemma keeled and 2-toothed at tip

awns   lemma awned from ⅓ below the tip, divergent (5–6 mm long), conspicuous, geniculate

glumes   unequal, first glume 1-nerved and shorter than lemma, second glume broader and 3-nerved

other   3 awns per spikelet, hence the name *Trisetum*

VEGETATIVE CHARACTERISTICS

culm   erect (15–50 cm tall), densely tufted, glabrous to downy pubescent

sheath   keeled at upper end, pubescent except on margins, scattered long hairs near collar

blade   flat to involute (3–12 cm long and 2–5 mm wide), tapering to a blunt point, distinctly veined, usually puberulent

ligule   membranous (2 mm long), rounded, erose

GROWTH CHARACTERISTICS   starts growth in the early spring, low production of viable seed, remains green until August

FORAGE VALUE   good for all classes of livestock and big game throughout the growing season and late in the fall

HABITAT   plains, alpine meadows, and slopes

Giant cane (*Arundinaria gigantea*)

Notes:

Tribe	BAMBUSEAE
Species	*Arundinaria gigantea* (Walt.) Muhl.
Common Name	Giant cane
Life Span	Perennial
Origin	Native
Season	Cool

INFLORESCENCE CHARACTERISTICS

type   panicles, panicle branchlets crowded toward the ends of the branches

spikelets   6- to 12-flowered (3–7 cm long)

awns   lemma sometimes tapers to an awn (4 mm long)

glumes   unequal, shorter than lemmas, pubescent, acuminate, sometimes wanting

VEGETATIVE CHARACTERISTICS

culm   erect (2–8 m tall), woody (bamboo), smooth

sheath   lower half as long as internodes; hairy, elevated collar; auricles with 10–12 stiff bristles

blade   on short petiole (petiole 1–2 mm long), wide, flat (15–40 cm long and 2–6 cm wide), margins serrulate, netted venation

ligule   membranous (1 mm long), firm

other   rhizomes

GROWTH CHARACTERISTICS   starts growth in March, flowers in April and May every 4–6 years, reproduces mainly from rhizomes

FORAGE VALUE   produces good forage for both cattle and wildlife and is grazed or browsed all year long

HABITAT   low, moist woodlands and along streams and swales

Broadleaf chasmanthium (*Chasmanthium latifolium*)

Tribe	CENTOTHECEAE
Species	*Chasmanthium latifolium* (Michx.) Yates
Common Name	Broadleaf chasmanthium (inland sea oats)
Life Span	Perennial
Origin	Native
Season	Cool

INFLORESCENCE CHARACTERISTICS

type   panicle (10–20 cm long), open, drooping

spikelets   8- to 15-flowered, (2–5 cm long), 1–6 empty lemmas above and below fertile lemmas, strongly compressed, keel ciliate with soft ascending hairs

glumes   do not exceed lower lemma (5–7 mm long)

other   pedicles capillary, tips of florets curve inward

VEGETATIVE CHARACTERISTICS

culm   erect (1.0–1.5 m tall)

sheath   glabrous and tightly clasping the internode

blade   flat (10–20 cm long and 1–2 cm wide), narrowly lanceolate

ligule   membranous (1 mm long or less), ciliate

other   short, strong rhizomes

GROWTH CHARACTERISTICS   starts growth in April, flowers June–October, reproduces by rhizomes and seeds

FORAGE VALUE   little forage value for domestic livestock or wildlife

HABITAT   stream banks and moist woodlands

Sideoats grama (*Bouteloua curtipendula*)

Tribe	CHLORIDEAE
Species	*Bouteloua curtipendula* (Michx.) Torr.
Common Name	Sideoats grama (banderilla, avenilla, banderita)
Life Span	Perennial
Origin	Native
Season	Warm

INFLORESCENCE CHARACTERISTICS

type   panicle (10–30 cm long) of 35–80 spicate primary branches (1–2 cm long)

spikelets   1 perfect floret per spikelet with imperfect floret reduced to 3 bristles; lemma 3-veined and 3-toothed at top; pendulous

awns   lemma sometimes with short awn (1–2 mm long)

glumes   unequal, tapering (0.6–1.0 cm long)

other   individual branches turned to one side of inflorescence, distant, branch base remains on culm after disarticulation

VEGETATIVE CHARACTERISTICS

culm   erect (0.2–1.0 m tall), tufted, smooth, purple at nodes

sheath   round, glabrous to somewhat pilose, prominently veined; collar pilose on margin

blade   flat to subinvolute (2–30 cm long and 2–4 mm wide), margins with scattered hairs from bulbous bases

ligule   membranous (0.5 mm long), ciliate

other   scaly rhizomes, leaves mostly basal

GROWTH CHARACTERISTICS   starts growth in early spring and flowers July to September; reproduces by seed, tillers, and rhizomes

FORAGE VALUE   good for all classes of livestock throughout summer and fall, remains moderately palatable into winter

HABITAT   plains, prairies, and rocky hills

Black grama (*Bouteloua eriopoda*)

Notes:

72

Tribe	CHLORIDEAE
Species	*Bouteloua eriopoda* (Torr.) Torr.
Common Name	Black grama (navajita negra)
Life Span	Perennial
Origin	Native
Season	Warm

INFLORESCENCE CHARACTERISTICS

type   panicle of 3–8 spicate primary branches; branches 2–5 cm long slender, delicate, white-lanate

spikelets   12–20 (0.7–1.0 cm long), not crowded, comb-like

awns   lemma awned (1.5–3.0 mm long)

glumes   unequal, pointed, glabrous

VEGETATIVE CHARACTERISTICS

culm   decumbent or stoloniferous, slender, arched internodes (20–60 cm long), woolly pubescent internodes, swollen and woolly base

sheath   glabrous

blade   mostly basal (0.5–2.0 mm wide), flexuous, pointed, soft, hairy above (adaxial)

ligule   ring of short hairs, rounded

other   stolons

GROWTH CHARACTERISTICS   starts growth when sufficient moisture is available in July, low seed viability, reproduction by stolons and tillers, frequently grows in almost pure stands

FORAGE VALUE   excellent for all classes of livestock and wildlife throughout the year

HABITAT   mesas, hills, and dry open ground

Blue grama (*Bouteloua gracilis*)

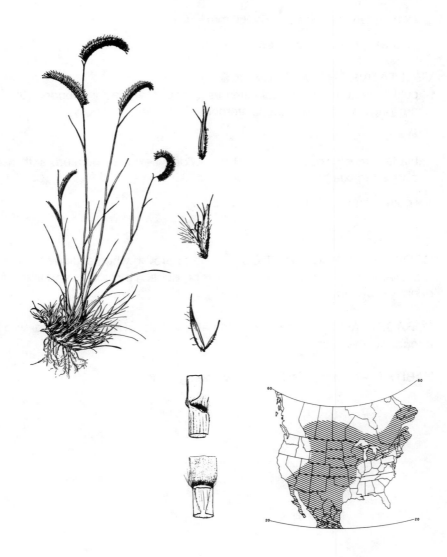

Tribe	CHLORIDEAE
Species	*Bouteloua gracilis* (H.B.K.) Lag. *ex* Steud.
Common Name	Blue grama (navajita azul)
Life Span	Perennial
Origin	Native
Season	Warm

INFLORESCENCE CHARACTERISTICS

type   panicle of 1–3 (sometimes 4) spicate primary branches; branches 1.5–5.0 cm long

spikelets   1 perfect floret, 1 or more reduced florets, 40–90 spikelets per branch

awns   lemmas awned, generally 3 (1–3 mm long)

glumes   1-nerved, hairless or swollen-based hairs on the nerve

other   rachis not projecting beyond spikelets

VEGETATIVE CHARACTERISTICS

culm   erect (20–60 cm tall), densely tufted

sheath   rounded, glabrous to pilose, long-haired at junction with blade

blade   tapering (5–25 cm long and 3–5 mm wide), flat or loosely involute

ligule   few soft hairs (0.5 mm long)

GROWTH CHARACTERISTICS   starts growth in May or June, matures in 2 months, dormant during dry periods in the summer

FORAGE VALUE   good for all classes of livestock and wildlife

HABITAT   open plains and rocky slopes

Hairy grama (*Bouteloua hirsuta*)

Tribe	CHLORIDEAE
Species	*Bouteloua hirsuta* Lag.
Common Name	Hairy grama (navajita vellosa)
Life Span	Perennial
Origin	Native
Season	Warm

INFLORESCENCE CHARACTERISTICS

type   panicle of 2 (1–4) spicate primary branches (branches 2.5–3.5 cm long), comb-like, dark glands

spikelets   20–50 spikelets (6 mm long), 1 perfect and 1–3 reduced florets

awns   lower rudimentary floret with 3 awns (4 mm long), glumes minutely awned at the tip

glumes   unequal, first short, second longer (3–5 mm long) and tuberculate-hirsute

other   rachis extending beyond spikelets (5–6 mm)

VEGETATIVE CHARACTERISTICS

culm   tufted (15–60 cm tall), 4–8 nodes

sheath   veined, hairless, or lowermost thinly soft haired; collar hairy with glandular marginal hairs

blade   flat or involute (3–10 cm long and 1–3 mm wide), narrow, pointed, marginal glandular hairs

ligule   membranous (0.25 mm long), truncate

GROWTH CHARACTERISTICS   starts growth by mid-July or when moisture is available near that time period

FORAGE VALUE   fair for both livestock and wildlife, palatability is highest late in the growing season

HABITAT   rough, rocky ridges; loose sands; dry, shallow uplands

Slender grama (*Bouteloua repens*)

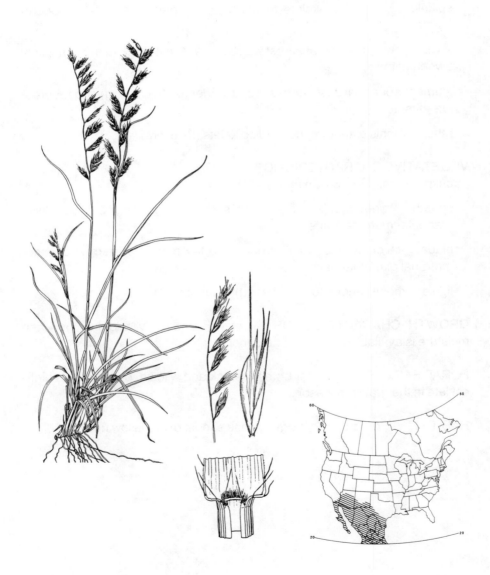

Tribe	CHLORIDEAE
Species	*Bouteloua repens* (H.B.K.) Scribn. & Merr.
Common Name	Slender grama (navajita)
Life Span	Perennial
Origin	Native
Season	Warm

INFLORESCENCE CHARACTERISTICS

type   panicle of 4–9 spicate primary branches (branches 3–7 cm long)

spikelets   generally 4–7 spikelets per branch (1–2 cm long excluding awns)

awns   lemma of rudiment with 3 distinct awns (3–7 mm long)

glumes   slightly unequal (3–6 mm long), scabrous on the strong mid-nerve

VEGETATIVE CHARACTERISTICS

culm   tufted (20–45 cm tall), slender, weak

sheath   round

blade   thin, rolled (5–16 cm long and 1–3 mm wide), sparsely ciliate on margins, mostly basal

ligule   membranous, minute (less than 0.3 mm long), fringed

GROWTH CHARACTERISTICS   starts growth in April when adequate moisture is available, reproduces by seeds and tillers

FORAGE VALUE   good for livestock and wildlife, especially valuable in fall, winter, and spring

HABITAT   open or brushy pastures, along streambanks, and road rights-of-way

Buffalograss (*Buchloe dactyloides*)

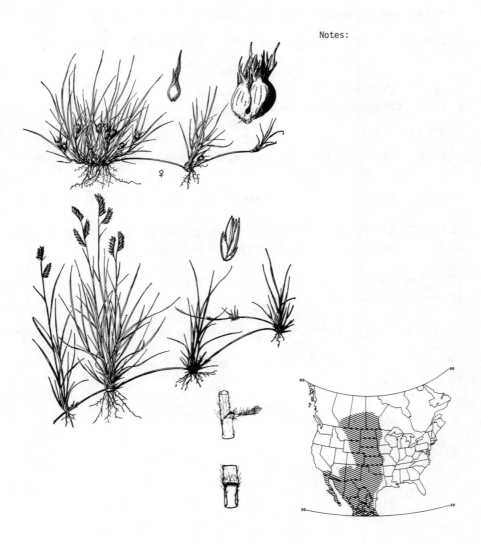

Tribe	CHLORIDEAE
Species	*Buchloe dactyloides* (Nutt.) Engelm.
Common Name	Buffalograss (zacate chino, bufalo)
Life Span	Perennial
Origin	Native
Season	Warm

INFLORESCENCE CHARACTERISTICS

type   panicle of 1–2 spicate primary branches (male plants), bur-like cluster of spikelets (female plants), dioecious

spikelets   male spikelets 6–12 flowered (4.0–5.5 mm long); female spikelets 1-flowered in bur-like clusters of 3–7

glumes   glumes of female and male spikelets unequal

other   female inflorescence (bur) usually shorter than leaves, male inflorescence (flag) exceeds leaves

VEGETATIVE CHARACTERISTICS

culm   staminate culm erect (5–25 cm tall), nodes glabrous

sheath   rounded on the back, hairless except for a few marginal hairs near the collar

blade   curly (2–15 cm long and 1.0–2.5 mm wide), flat, sparsely pilose on one or both surfaces

ligule   fringe of hairs (0.5–1.0 mm long)

other   stolons

GROWTH CHARACTERISTICS   starts growth when temperature is about 20° C, grows when adequate moisture is available, seed matures in about 6 weeks

FORAGE VALUE   good for all classes of livestock and fair for wildlife

HABITAT   dry plains, on medium to fine-textured soils

Hooded windmillgrass (*Chloris cucullata*)

Tribe	CHLORIDEAE
Species	*Chloris cucullata* Bisch.
Common Name	Hooded windmillgrass
Life Span	Perennial
Origin	Native
Season	Warm

INFLORESCENCE CHARACTERISTICS

type   panicle of digitate branches, numerous, radiating (2–5 cm long), curled

spikelets   crowded, triangular, inflated

awns   sterile floret may be awnless or with an awn (to 1.5 mm long); glumes awn tipped

glumes   lanceolate to obovate, first glume shorter (0.5–0.7 mm long), second glume longer (1.0–1.5 mm long)

other   inflorescence brown at maturity

VEGETATIVE CHARACTERISTICS

culm   erect (15–60 cm tall), tufted, flat base

sheath   flat, glabrous, white margins

blade   folded, blunt-tipped (5–20 cm long and 2–4 mm wide), glabrous to scabrous, white midrib

ligule   membranous (1–2 mm long), rounded, ciliate

other   occasionally with short rhizomes

GROWTH CHARACTERISTICS   starts growth in early spring, may produce several inflorescences per year

FORAGE VALUE   fair for livestock and wildlife

HABITAT   plains and sandy soils

Multiflowered false-rhodesgrass (*Chloris pluriflora*)

Tribe	CHLORIDEAE
Species	*Chloris pluriflora* (Fourn.) Clayton
Common Name	Multiflowered false-rhodesgrass (zacate pelillo)
Life Span	Perennial
Origin	Native
Season	Warm

INFLORESCENCE CHARACTERISTICS

type   panicle of 7–20 spicate primary branches (branches 7–20 cm long), inflorescence terminal

spikelets   overlapping (5–9 mm long), 1 or 2 perfect florets, 2 or 3 sterile florets

awns   lemma of lowermost floret with 3 awns, unequal (center awn 0.8–1.5 cm and lateral awns 0.5–1.5 mm long)

glumes   lanceolate, glabrous except for the scabrous midrib, unequal, first glume 2–3 mm long, second glume 3–5 mm long

other   dense and feathery inflorescence

VEGETATIVE CHARACTERISTICS

culm   erect (0.5–1.5 m tall), glabrous

sheath   glabrous to sparsely hirsute

blade   scabrous to sparsely hirsute (10–30 cm long and 0.5–1.0 cm wide)

ligule   membranous with a fringe of hairs (2 mm long)

other   may be stoloniferous

GROWTH CHARACTERISTICS   growth determined by moisture rather than temperature, starts growth in late spring, several seed crops produced per year

FORAGE VALUE   good to excellent for livestock, occasional deer use

HABITAT   plains and dry woods on sandy loam soils

Bermudagrass (*Cynodon dactylon*)

Notes:

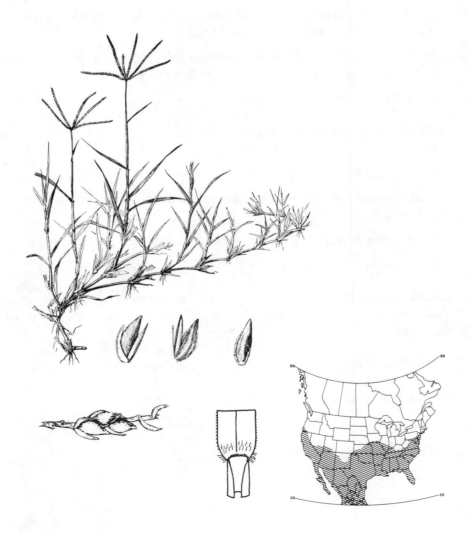

Tribe	CHLORIDEAE
Species	*Cynodon dactylon* (L.) Pers.
Common Name	Bermudagrass (bermuda, pata de gallo, agrarista)
Life Span	Perennial
Origin	Introduced (from Africa)
Season	Warm

INFLORESCENCE CHARACTERISTICS

type   panicle of 2–7 digitate branches (branches 2–6 cm long)

spikelets   sessile, 2 rows on one side of the branch, not inflated; 1 perfect floret; lemma boat-shaped (2.0–2.5 mm long)

glumes   lanceolate, subequal, 1-nerved, ⅔ as long as lemma

other   floriferous to the base of the spicate branches

VEGETATIVE CHARACTERISTICS

culm   flattened, low (10–50 cm tall), weak, mat-forming

sheath   rounded, glabrous except for tufts of hair on either side of the collar

blade   linear, flat, or folded (3–12 cm long and 1–3 mm wide); pilose on upper surface

ligule   membranous (0.2–0.5 mm long), conspicuous ring of white hairs at margins

other   scaly rhizomes or strong and flat stolons

GROWTH CHARACTERISTICS   high salt tolerance, reproduces by seeds, rhizomes, and stolons

FORAGE VALUE   good for cattle and poor for wildlife; used for pasture in spring and early summer

HABITAT   open ground, fields, waste places, planted in pastures

Curly mesquite (*Hilaria belangeri*)

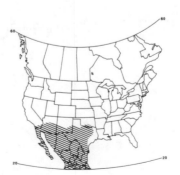

Tribe	CHLORIDEAE
Species	*Hilaria belangeri* (Steud.) Nash
Common Name	Curly mesquite (toboso menudo)
Life Span	Perennial
Origin	Native
Season	Warm

INFLORESCENCE CHARACTERISTICS

type   spike (2.0−3.5 cm long), 4−8 inflorescence nodes

spikelets   3 spikelets per node (4.5−6.0 mm long), fan-shaped

awns   midrib of glumes extended into short awn

glumes   glumes of lateral spikelets scabrous, united below, usually shorter than lemmas, outer glume broadened above; glumes of central spikelet subequal, glabrous or scabrous

other   rachis axis flat and strongly angled at each node

VEGETATIVE CHARACTERISTICS

culm   erect (10−30 cm tall), villous at nodes, internodes alternately curved

sheath   glabrous

blade   flat or less commonly involute (5−20 cm long and 1−2 mm wide), scabrous, pilose (1−2 mm long), hairy-glandular

ligule   membranous (0.5−1.5 mm long), erose

other   slender stolons, internodes wiry, nodes pubescent

GROWTH CHARACTERISTICS   starts growth in late spring, seedheads emerge 1 month or more later, highly drought resistant; produces a relatively large amount of forage for the plant size

FORAGE VALUE   fair for cattle, sheep, goats, deer, and antelope; cures well

HABITAT   slopes, dry hillsides, and grassy or brushy plains; rocky soils

Galleta (*Hilaria jamesii*)

Tribe	CHLORIDEAE
Species	*Hilaria jamesii* (Torr.) Benth.
Common Name	Galleta
Life Span	Perennial
Origin	Native
Season	Warm

INFLORESCENCE CHARACTERISTICS

type   spike (3–8 cm long), cluster of 3 spikelets (sessile) at each node

spikelets   center spikelet perfect (6–8 mm long), lateral spikelets staminate and 2-flowered

awns   first glume of lateral spikelets awned from back (3–5 mm long)

glumes   glumes of lateral spikelets acute

other   spikelets bearded at base with long hairs, not fan-shaped, rachis wavy

VEGETATIVE CHARACTERISTICS

culm   erect (30–50 cm tall), base may be decumbent, nodes villous

sheath   veined, collar glabrous, lacking auricles

blade   flat at base (3–12 cm long and 2–5 mm wide), upper ⅔ often rolled

ligule   membranous (1–3 mm long), truncate, lanceolate, erose

other   scaly rhizomes

GROWTH CHARACTERISTICS   grows from rhizomes and seed; growth mainly in summer after sufficient rain

FORAGE VALUE   good for all classes of livestock and wildlife while it is green, poor to fair during the dormant periods

HABITAT   deserts, canyons, and dry plains; heavy-textured soils

Tobosa (*Hilaria mutica*)

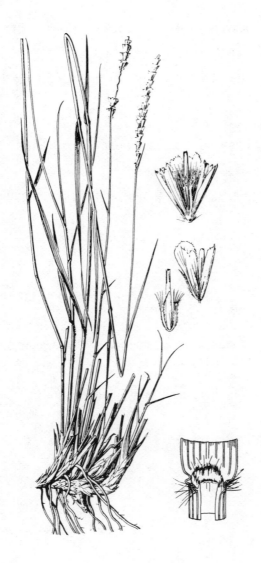

92

Tribe	CHLORIDEAE
Species	*Hilaria mutica* (Buckl.) Benth.
Common Name	Tobosa (toboso)
Life Span	Perennial
Origin	Native
Season	Warm

INFLORESCENCE CHARACTERISTICS

type   spike (4−6 cm long), wavy rachis, 3 spikelets per node

spikelets   in clusters of 3 (6−9 mm long), sessile, fan-shaped, occasionally bearded at the base with short hairs (2−3 mm long), lateral spikelets staminate

awns   inner glumes of lateral spikelets with a rough or hairy awn (0.5−3.0 mm long), glumes of center spikelet awn-tipped

glumes   glumes of lateral spikelets broadened, glumes of center spikelet narrow and short

VEGETATIVE CHARACTERISTICS

culm   bases decumbent (30−75 cm tall), wiry, lower nodes pubescent, upper nodes glabrous, internodes glabrous

sheath   glabrous, veined, collar hairy on margin, lacking auricles

blade   flat to rolled (5−10 cm long and 2−4 mm wide), veined, glabrous (occasionally hairy adaxially)

ligule   membranous (1−2 mm long), truncate, ciliate

other   rhizomes

GROWTH CHARACTERISTICS   grows vigorously from stout rhizomes, low seed production, growth starts in late spring or summer after sufficient rain

FORAGE VALUE   good to fair for cattle and horses, fair for sheep, poor for wildlife, becomes relatively unpalatable when mature

HABITAT   dry, rocky slopes; dry upland plains and plateaus; heavy clay soils

Tumblegrass (*Schedonnardus paniculatus*)

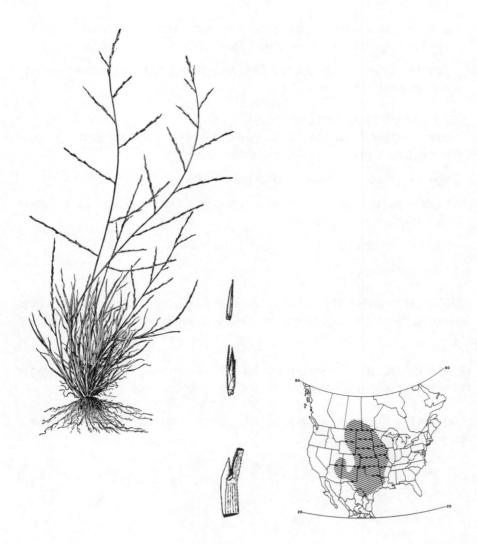

Tribe	CHLORIDEAE
Species	*Schedonnardus paniculatus* (Nutt.) Trel.
Common Name	Tumblegrass (Texas crabgrass)
Life Span	Perennial
Origin	Native
Season	Warm

INFLORESCENCE CHARACTERISTICS

type   panicle of several spicate primary branches (30–60 cm long), few branches (2–20 cm long), curved at maturity

spikelets   1-flowered, slender, sessile (3–4 mm long), imbedded in branch

glumes   unequal, second glume as long as lemma, lanceolate, 1-nerved

other   inflorescence breaks off at the base and tumbles in the wind

VEGETATIVE CHARACTERISTICS

culm   stiffly curving, erect-ascending (10–70 cm tall), tufted, often decumbent

sheath   laterally compressed and keeled

blade   mainly basal, twisted (2–12 cm long and 1–2 mm wide), glabrous, midrib prominent dorsally, scabrous on margins, white margins

ligule   membranous (2–3 mm long), rounded

GROWTH CHARACTERISTICS   grows from early spring to late fall when moisture is available

FORAGE VALUE   develops little herbage and has little value for livestock or wildlife

HABITAT   prairies, plains, and waste places

Marshhay cordgrass (*Spartina patens*)

Tribe	CHLORIDEAE
Species	*Spartina patens* (Ait.) Muhl.
Common Name	Marshhay cordgrass (saltmeadow cordgrass)
Life Span	Perennial
Origin	Native
Season	Warm

INFLORESCENCE CHARACTERISTICS

type   panicle of 2−7 spicate primary branches (each branch 3−8 cm long)

spikelets   1-flowered (0.7−1.2 cm long), palea slightly longer than lemma

glumes   unequal, first glume ½ spikelet length or less, second glume as long as spikelet, scabrous on nerve, glume margins slightly scabrous

VEGETATIVE CHARACTERISTICS

culm   single or in clusters (0.5−1.5 m tall)

sheath   rounded, glabrous

blade   narrow, tightly involute after drying (1−4 mm wide)

ligule   fringe of hairs, minute

other   may have creeping, fine or large rhizomes

GROWTH CHARACTERISTICS   growth begins in late spring, rapid growth in summer, seeds produced by October

FORAGE VALUE   good grazing for cattle, but poor for wildlife, good for muskrats, managed for winter grazing

HABITAT   salt marshes and sandy meadows along the coast

Prairie cordgrass (*Spartina pectinata*)

Notes:

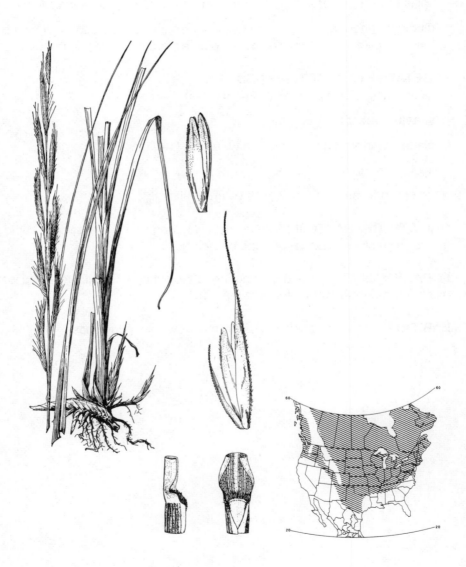

Tribe	CHLORIDEAE
Species	*Spartina pectinata* Link
Common Name	Prairie cordgrass (tall marshgrass, sloughgrass)
Life Span	Perennial
Origin	Native
Season	Warm

INFLORESCENCE CHARACTERISTICS

type   panicle of 6−40 spicate primary branches (each branch 4−15 cm long)

spikelets   1-flowered

awns   second glume awned (0.4−1.0 cm long), stout

glumes   unequal (first glume 6−7 mm long and second 8−9 mm long, excluding the awn), hispid-scabrous on keel, serrate margins

VEGETATIVE CHARACTERISTICS

culm   solitary or in small clusters (1.0−2.5 m tall), robust

sheath   open, distinctly veined, usually pubescent only in the throat

blade   flat when green (20−80 cm long and 0.6−1.5 cm wide), involute when dry, serrate margins, tapering to a point, scabrous

ligule   ring of hairs (1−3 mm long)

other   stout, scaly, widely-spreading rhizomes

GROWTH CHARACTERISTICS   growth starts in early spring, reproduces by seed and rhizomes

FORAGE VALUE   herbage is coarse and furnishes poor to fair forage for cattle, becomes unpalatable with maturity

HABITAT   marshy meadows, along swales, and ditches

California oatgrass (*Danthonia californica*)

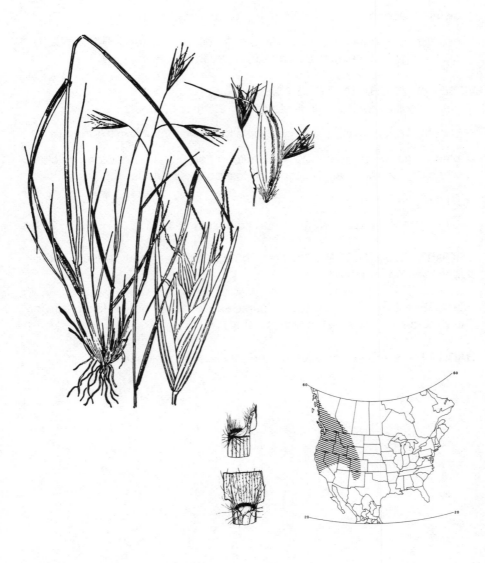

Tribe	DANTHONIEAE
Species	*Danthonia californica* Boland
Common Name	California oatgrass
Life Span	Perennial
Origin	Native
Season	Cool

INFLORESCENCE CHARACTERISTICS

type   panicle, spreading, open, often with 3 branches

spikelets   large (2 cm long), few (1–5), often divergent at right angle to culm, 3- to 6-flowered, florets small

awns   lemma awns arise apically between 2 teeth, geniculate (terminal segment 0.5–1.0 cm long)

glumes   equal (1.5–2.0 cm long), spreading

other   prominent swollen area in axils at base of pedicels, pedicels pubescent (1–2 cm long)

VEGETATIVE CHARACTERISTICS

culm   erect (0.3–1.0 m tall), glabrous

sheath   round, hyaline margins, prominently veined, collar pilose, pilose at throat

blade   flat to involute (5–25 cm long and 2–4 mm wide), upper blades often at right angles to culm, pubescent adaxially

ligule   membranous with a fringe of hairs (0.5–1.5 mm long)

GROWTH CHARACTERISTICS   starts growth in the early spring, seed matures by July; reproduces by seeds and tillers

FORAGE VALUE   good to excellent for all classes of livestock and wildlife

HABITAT   meadows, open woods, coastal prairies

Timber oatgrass (*Danthonia intermedia*)

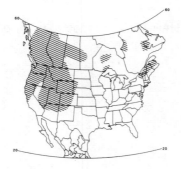

Tribe	DANTHONIEAE
Species	*Danthonia intermedia* Vasey
Common Name	Timber oatgrass (timber danthonia)
Life Span	Perennial
Origin	Native
Season	Cool

INFLORESCENCE CHARACTERISTICS

type   panicle (2–6 cm long), narrow, ascending branches

spikelets   usually 2 (may be several) large florets per spikelet (at least 1 cm long), lemma tipped with acuminate teeth

awns   lemma dorsally awned (up to 1 cm long), twisted

glumes   subequal (1.3–1.8 cm long), longer than lowermost floret, aristate-tipped

other   inflorescence often purplish in color, pedicels glabrous

VEGETATIVE CHARACTERISTICS

culm   tufted (10–50 cm tall), glabrous

sheath   glabrous (rarely pilose), pilose at the throat and usually on the collar

blade   mainly basal (5–25 cm long and 2–4 mm wide), flat or involute, ascending if on upper culm, often pubescent abaxially

ligule   membranous with a fringe of hairs (0.3–1.0 mm long)

GROWTH CHARACTERISTICS   starts growth in early spring; seeds mature by September; reproduces by seeds and tillers; cleistogamous spikelets in the lower sheaths

FORAGE VALUE   good to excellent for domestic livestock, deer, and elk; utilized in the spring

HABITAT   meadows, bogs, and as understory in forests

Parry oatgrass (*Danthonia parryi*)

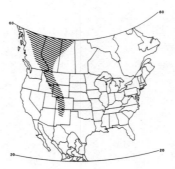

Tribe	DANTHONIEAE
Species	*Danthonia parryi* Scribn.
Common Name	Parry oatgrass (Parry danthonia)
Life Span	Perennial
Origin	Native
Season	Cool

INFLORESCENCE CHARACTERISTICS

type   panicle (3–7 cm long); branches ascending or appressed, more or less pubescent, lowermost 1–2 cm long, usually with 3–8 spikelets

spikelets   several flowered; lemma 11-nerved, bifid apex with long acuminate teeth and pilose hair on the back and margins (1.0 cm long); palea narrowed above, nearly as long as the lemma

awns   lemma awn geniculate, from between the cleft, terminal segment 0.8–1.2 cm long

glumes   equal (1.6–2.2 cm long), longer than the lowermost floret

VEGETATIVE CHARACTERISTICS

culm   stout (30–60 cm tall), dense clumps, somewhat enlarged at the base due to numerous overlapping sheaths

sheath   glabrous, pilose at the throat; a line or ridge on the collar, may be either glabrous or pubescent; cleistogamous spikelets may be found in axils of the sheath

blade   erect-flexuous (15–25 cm long), narrow or filiform, flat or involute, glabrous

ligule   a ring of hairs (about 1 mm long)

GROWTH CHARACTERISTICS   starts growth in spring, seed matures in July and August, reproduces by seeds and tillers; cleistogamous spikelets in lower sheatus

FORAGE VALUE   fair for cattle, seldom abundant

HABITAT   open grassland, open woods, and rocky slopes

Pine dropseed (*Blepharoneuron tricholepis*)

Tribe	ERAGROSTEAE
Species	*Blepharoneuron tricholepis* (Torr.) Nash
Common Name	Pine dropseed (pastillo del pinar)
Life Span	Perennial
Origin	Native
Season	Warm

INFLORESCENCE CHARACTERISTICS

type   panicle (5−20 cm long and 2−5 cm wide), long-exserted from the sheath, twisted, capillary branches, elliptical

spikelets   1-flowered (2.5−4.0 mm long), distinctive green-gray color, lemma with 3 villous nerves

glumes   broad, blunt, round, glabrous, first glume faintly 5-nerved and equal to or shorter than the faintly 3-nerved second glume, glumes shorter than lemma

VEGETATIVE CHARACTERISTICS

culm   erect (20−70 cm tall), tufted, nodes glabrous

sheath   glabrous, rounded on back

blade   basal leaves enrolled (5−20 cm long and 2−4 mm wide), glabrous or scabrous, margin scabrous

ligule   membranous (0.3−0.5 mm long), ciliate, rounded

GROWTH CHARACTERISTICS   starts growth in late spring or early summer, reproduces by seeds and tillers

FORAGE VALUE   palatability and quality of young plants is very good for all classes of livestock, stems are neglected or only slightly grazed after maturity

HABITAT   rocky, open slopes, and in dry woodlands at medium to high elevations

Prairie sandreed (*Calamovilfa longifolia*)

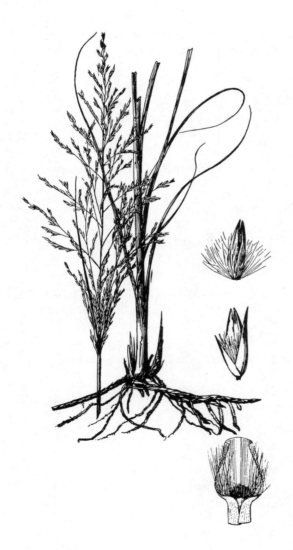

Tribe	ERAGROSTEAE
Species	*Calamovilfa longifolia* (Hook.) Scribn.
Common Name	Prairie sandreed
Life Span	Perennial
Origin	Native
Season	Warm

INFLORESCENCE CHARACTERISTICS

type   panicle (15–35 cm long), narrow, shiny

spikelets   1-flowered (6–7 mm long), callus hairs copious; lemma ciliate, 1-nerved

glumes   boat-shaped, glabrous, first shorter than the second

VEGETATIVE CHARACTERISTICS

culm   erect (0.5–1.8 m tall), solitary, robust

sheath   glabrous to pubescent, distinctly veined, short hair on margin, hair at the throat (2–3 mm long)

blade   tapered (10–50 cm long and 0.3–1.2 cm wide) involute, basal portion keeled

ligule   ring of hairs (0.5–3.0 mm long)

other   stout, scaly rhizomes; inflated collars

GROWTH CHARACTERISTICS   grows rapidly in late spring and throughout the summer, remains green until frost, drought tolerant, reproduces by seeds and rhizomes

FORAGE VALUE   fair for cattle throughout the summer, moderate use by wildlife; it cures well and provides good standing winter feed, quality of hay may be relatively poor

HABITAT   sand hills, sandy prairies, and open woods

Weeping lovegrass (*Eragrostis curvula*)

Notes:

Tribe	ERAGROSTEAE
Species	*Eragrostis curvula* (Schrad.) Nees
Common Name	Weeping lovegrass (zacate horon)
Life Span	Perennial
Origin	Introduced (from South Africa)
Season	Warm

INFLORESCENCE CHARACTERISTICS

type panicle (25–40 cm long and 8–12 cm wide), open

spikelets 6- to 12-flowered, gray-green color, short pediceled; lemma obtuse (2.5 mm long), nerves prominent, 3-nerved

glumes unequal (first 1.8 mm long and second 2.8 mm long), membranous, sharp-pointed

other panicle branches solitary or in pairs, naked at the base, densely pilose in the axils

VEGETATIVE CHARACTERISTICS

culm tufted (0.6–1.5 m tall)

sheath shorter than internodes, narrow, keeled, hairy toward base, not pilose at the summit, but with long, soft hairs at the throat

blade elongate (stem leaves 20–30 cm long and 1.0–1.5 mm wide, basal leaves much longer), involute

ligule ring of hairs (0.5–1.0 mm long), white

GROWTH CHARACTERISTICS starts growth in late spring, reproduces by seeds and tillers, drought resistant

FORAGE VALUE fair for livestock but relatively poor for wildlife, less palatable than many other seeded species

HABITAT planted as a forage grass, persistent on roadsides and in sandy fields and waste areas

Sand lovegrass (*Eragrostis trichodes*)

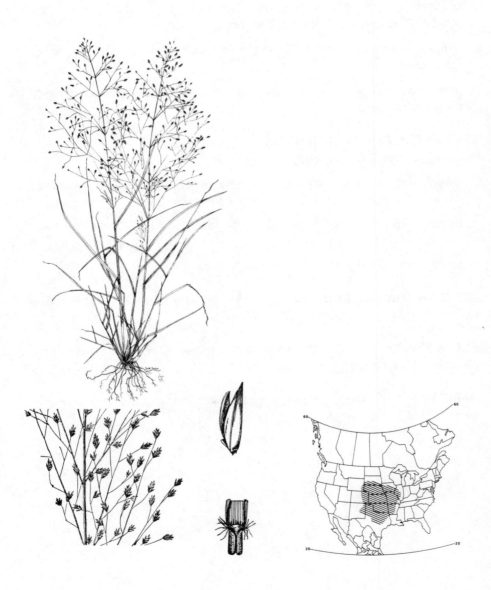

Tribe	ERAGROSTEAE
Species	*Eragrostis trichodes* (Nutt.) Wood
Common Name	Sand lovegrass
Life Span	Perennial
Origin	Native
Season	Warm

INFLORESCENCE CHARACTERISTICS

type   panicle (35–55 cm long and 7–30 cm wide), diffuse, may be ½ the length of the plant, capillary branches

spikelets   4- to 18-flowered (0.4–1.0 cm long), long pediceled, lemmas 3-nerved, lateral nerves strong

glumes   thin, pointed

other   panicle purple or red, panicle branches in groups of 3 or 4, sparsely pilose in the axils

VEGETATIVE CHARACTERISTICS

culm   erect (0.4–1.8 m tall), tufted

sheath   pilose at summit, occasionally hairy on the back or margins

blade   elongate, flat (15–40 cm long and 2–8 mm wide), scabrous on upper surface

ligule   ring of hairs (0.2–0.5 mm long)

GROWTH CHARACTERISTICS   starts growth as much as 2 weeks earlier than other warm season grasses, remains green into the fall, reproduces by seeds and tillers

FORAGE VALUE   excellent for all classes of livestock and wildlife during summer, fair to good after maturity, cures well

HABITAT   deep sand, sandy loam

Green sprangletop (*Leptochloa dubia*)

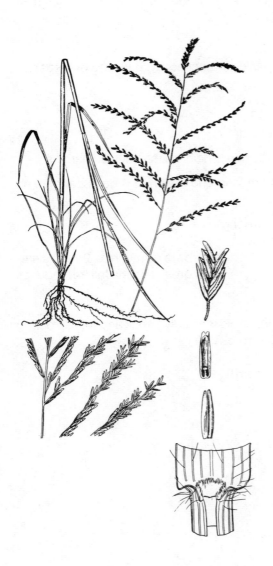

Tribe	ERAGROSTEAE
Species	*Leptochloa dubia* (H.B.K.) Nees
Common Name	Green sprangletop (zacate gigante)
Life Span	Perennial
Origin	Native
Season	Warm

INFLORESCENCE CHARACTERISTICS

type   panicle of racemose branches, 2–15 branches (4–12 cm long), alternate, loosely erect or spreading

spikelets   2- to 8-flowered, nearly sessile, loosely to closely overlapping; lemmas broad, notched at apex, 3-nerved

glumes   unequal (second glume 4–5 mm long, first shorter) lanceolate, translucent with green nerves

VEGETATIVE CHARACTERISTICS

culm   tufted (0.3–1.1 m tall), erect, unbranched above the base, nodes dark brown to black

sheath   keeled, flattened, glabrous or lower ones pilose, collar with hairy margin (3–5 mm long), basal sheaths with cleistogamous spikelets

blade   flat or involute on drying (15–25 cm long and 4–5 mm wide), sometimes drooping, midrib prominent above

ligule   membranous (0.5 mm long), fringed

GROWTH CHARACTERISTICS   starts growth in April, continued growth depends on available moisture, reproduces by seeds and tillers

FORAGE VALUE   good for livestock and fair for wildlife

HABITAT   rocky hills, canyons, and sandy soil

Mountain muhly (*Muhlenbergia montana*)

Tribe	ERAGROSTEAE
Species	*Muhlenbergia montana* (Nutt.) Hitchc.
Common Name	Mountain muhly
Life Span	Perennial
Origin	Native
Season	Warm

INFLORESCENCE CHARACTERISTICS

type   panicle (5–20 cm long and 1–3 cm wide), narrow

spikelets   1-flowered; lemma (3.0–4.5 mm long), greenish to grayish with dark green or purple blotches or bands, with fine hairs on nerves

awns   lemma awned from tip, awn (0.6–2.5 cm) flexuous, straight or bent

glumes   thin, scabrous to nearly glabrous, second glume 3-toothed, first glume acute and shorter than second glume

other   rather loosely flowered, primary panicle branches usually floriferous to within 1 cm of base

VEGETATIVE CHARACTERISTICS

culm   densely tufted (15–80 cm tall), erect, stout

sheath   glabrous, rounded, lower ones often becoming flat and spreading

blade   flat to involute (5–25 cm long and 1–3 mm wide), pointed

ligule   membranous (0.4–1.8 cm long), hyaline, pointed

GROWTH CHARACTERISTICS   starts growth in late spring, matures August–September, reproduces by seeds and tillers

FORAGE VALUE   fair to good palatability for livestock while immature

HABITAT   canyons, mesas, and rocky hills

Bush muhly (*Muhlenbergia porteri*)

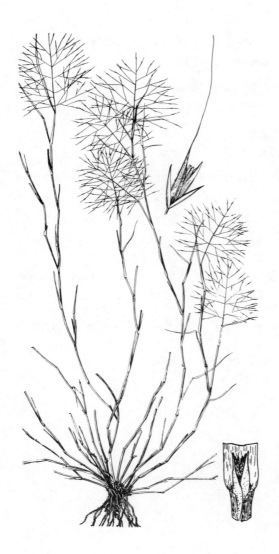

Tribe	ERAGROSTEAE
Species	*Muhlenbergia porteri* Scribn. *ex* Beal
Common Name	Bush muhly (zacate araña, zacate matorralero)
Life Span	Perennial
Origin	Native
Season	Warm

INFLORESCENCE CHARACTERISTICS

type   panicle (5–10 cm long), open, much branched, branches slender, purple, nearly as broad as long, breaks at base

spikelets   1-flowered, widely spaced on long pedicels (0.5–2.0 cm long including the awn), lemma sparsely pubescent

awns   lemma awned (0.5–1.3 cm long), delicate

glumes   slightly unequal (2–3 mm long), narrow, acuminate, thin, glabrous

VEGETATIVE CHARACTERISTICS

culm   spreading to ascending (0.3–1.0 m tall), wiry, geniculate, woody and knotty at the base, branching from the nodes

sheath   open, spreading away from the internodes, mostly shorter than the internodes

blade   thin, flat, becoming involute (2–8 cm long and 0.5–2.0 mm wide), pointed

ligule   membranous (1–2 mm long), truncate, lacerate

other   branching rhizomes

GROWTH CHARACTERISTICS   growth starts at nodes and crown each year, stems do not die back each year, drought resistant

FORAGE VALUE   excellent for all classes of livestock, remains green year-long if moisture is adequate which makes it especially palatable in winter

HABITAT   grows under the protection of various shrubs on dry mesas and hills, canyons, and rocky deserts

Green muhly (*Muhlenbergia racemosa*)

Tribe	ERAGROSTEAE
Species	*Muhlenbergia racemosa* (Michx.) B.S.P.
Common Name	Green muhly (marsh muhly)
Life Span	Perennial
Origin	Native
Season	Warm

INFLORESCENCE CHARACTERISTICS

type   panicle (3–15 cm long and 0.4–1.5 cm wide), tight, terminal, inter-rupted or lobed, densely flowered

spikelets   1-flowered; lemma acuminate, rarely with short awn, 3-nerved

awns   glumes awn-tipped (2 mm long), scabrous

glumes   about equal, awn-tipped (4.5–7.5 mm long including awn), 1-nerved

other   leaves up to base of inflorescence, inflorescence green to purple

VEGETATIVE CHARACTERISTICS

culm   erect (0.3–1.0 m tall), often branching at middle nodes

sheath   keeled, loose, glabrous or scabrous, distinctly veined, hyaline margins

blade   flat or loosely involute (4–18 cm long and 2–8 mm wide), erect, scabrous

ligule   membranous (0.5–1.5 mm long), erose, truncate to rounded

other   branching rhizomes

GROWTH CHARACTERISTICS   starts growth in late spring, flowers July to August

FORAGE VALUE   fair for cattle when immature, poor for wildlife

HABITAT   meadows, prairies, alluvial soil along rivers and ditches, rocky slopes, dry ground, and waste places

Ring muhly (*Muhlenbergia torreyi*)

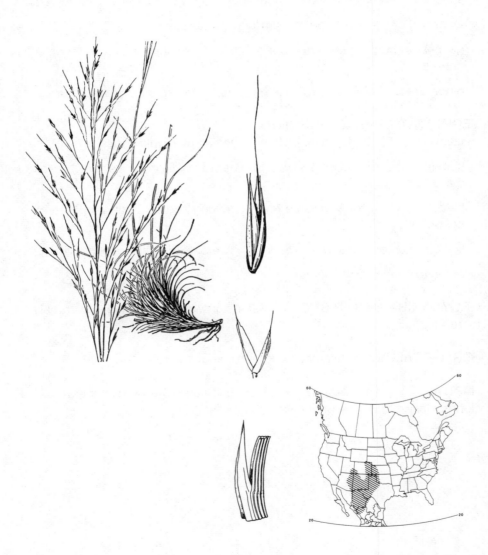

Tribe	ERAGROSTEAE
Species	*Muhlenbergia torreyi* (Kunth) Hitchc. *ex* Bush
Common Name	Ring muhly
Life Span	Perennial
Origin	Native
Season	Warm

INFLORESCENCE CHARACTERISTICS

type   panicle (7–25 cm long and 4–12 cm wide), diffuse, branchlets and pedicels appressed or at maturity spreading

spikelets   1-flowered (2.0–3.5 mm long), lemmas black

awns   lemma awned (1–3 mm long), hair-like; glumes awn-tipped

glumes   shorter than lemma, awn-tipped or irregulary toothed

other   pedicels equaling or longer than the spikelets

VEGETATIVE CHARACTERISTICS

culm   densely tufted (10–40 cm tall), decumbent base

sheath   oval, glabrous, broad hyaline margins

blade   arcuate (1–4 cm long and 0.5–2.0 mm wide), basal, involute, forming curly cushion, sharp-pointed

ligule   membranous (2–7 mm long), acuminate

other   short rhizomes, grows in a ring

GROWTH CHARACTERISTICS   starts growth in late spring or early summer, reproduces from seeds, tillers, and short rhizomes

FORAGE VALUE   fair to good for cattle when green, very low forage production

HABITAT   on clay soils of canyons and rocky slopes

Blowoutgrass (*Redfieldia flexuosa*)

Tribe	ERAGROSTEAE
Species	*Redfieldia flexuosa* (Thurb.) Vasey
Common Name	Blowoutgrass
Life Span	Perennial
Origin	Native
Season	Warm

INFLORESCENCE CHARACTERISTICS

type   panicle, open (⅓–½ as long as the culm), oblong; branches capillary and dichotomous

spikelets   1- to 6-flowered (5–8 mm long), V-shaped, lemma translucent, tuft of hair at base of floret

glumes   thin, narrow, lanceolate, slightly unequal to nearly equal (2–5 mm long), first glume 1-nerved, second glume 3-nerved

VEGETATIVE CHARACTERISTICS

culm   erect (0.5–1.0 m tall), coarse, tough, glabrous

sheath   nearly round, smooth, open, glabrous or short pubescent on raised veins, collar slightly expanded

blade   elongate (10–50 cm long and 1.5–8.0 mm wide), involute, tapering to a point, glabrous, flexuous

ligule   membranous (2–3 mm long), densely ciliate, rounded

other   long, slender rhizomes with fibrous roots

GROWTH CHARACTERISTICS   reproduces from seed and rhizomes, occurs in colonies, starts growth in spring, flowers July to October

FORAGE VALUE   fair for cattle in the summer, but is not readily grazed where other grasses are present

HABITAT   occurs on sandy soils that are subject to movement by wind

Burrograss (*Scleropogon brevifolius*)

Notes:

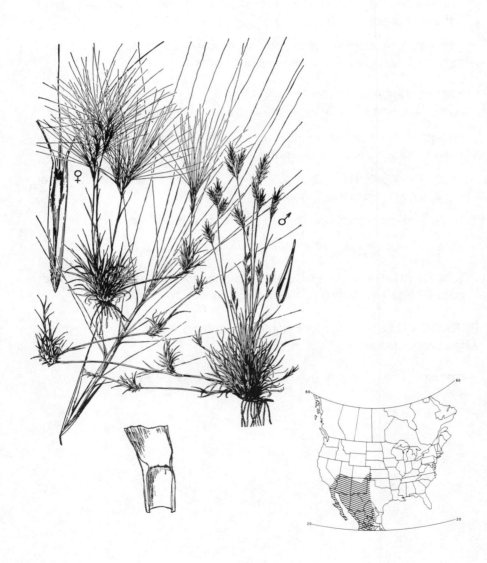

126

Tribe	ERAGROSTEAE
Species	*Scleropogon brevifolius* Phil.
Common Name	Burrograss (zacate burro, burrero)
Life Span	Perennial
Origin	Native
Season	Warm

INFLORESCENCE CHARACTERISTICS

type   contracted panicle or spicate racemes (1–5 cm long, excluding awns)

spikelets   unisexual, plants dioecious, sexes dissimilar, staminate spikelets 5- to 20-flowered (2–3 cm long), pistillate spikelets 3- to 5-flowered (2.5–3.0 cm long)

awns   lemma of pistillate spikelet with 3 awns (5–10 cm long), twisted

glumes   glumes of pistillate spikelets unequal, lanceolate, 3-nerved; glumes of staminate spikelets thin, pale, lanceolate, separated by short internode

VEGETATIVE CHARACTERISTICS

culm   tufted (10–25 cm tall)

sheath   short, strong-nerved, upper sheaths glabrous, lower sheaths hispid or villous

blade   basal, flat or folded (2–8 cm long and 1–2 mm wide), sharp-pointed or yucca-like, twisted

ligule   minute fringe of hairs

other   wiry, creeping stolons (internodes 5–15 cm long)

GROWTH CHARACTERISTICS   starts growth in May or June, inflorescences appear in 1 month, reproduces from stolons and seeds

FORAGE VALUE   poor for livestock and wildlife

HABITAT   soil of deteriorated sites that formerly supported tobosa, generally on clay, semiarid plains and open valleys

Alkali sacaton (*Sporobolus airoides*)

Tribe	ERAGROSTEAE
Species	*Sporobolus airoides* (Torr.) Torr.
Common Name	Alkali sacaton (zacatón alcalino)
Life Span	Perennial
Origin	Native
Season	Warm

INFLORESCENCE CHARACTERISTICS

type   panicle (20–45 cm long and 15–25 cm wide), pyramidal, open, usually not enclosed in sheath

spikelets   1-flowered (1.3–2.8 mm long), mostly on spreading pedicels (0.5–2.0 mm long), branchlets naked at the base, spikelets distinct and not overlapped

glumes   unequal, first glume ½ as long as lemma, second glume as long as lemma

VEGETATIVE CHARACTERISTICS

culm   erect (0.5–1.5 m tall), hard or woody base, bleached base, glabrous, shining

sheath   rounded, collar glabrous to sparsely hairy (2–4 mm long), ciliate margins

blade   flat or becoming involute (5–45 cm long and 2–6 mm wide), pointed, wide at base

ligule   membranous, ciliate (0.5 mm long with hairs 1–3 mm long)

GROWTH CHARACTERISTICS   reproduces from seeds and tillers, seeds remain viable for many years

FORAGE VALUE   fair for livestock and poor for wildlife

HABITAT   moderately moist alkaline soils in meadows and valleys

129

Tall dropseed (*Sporobolus asper*)

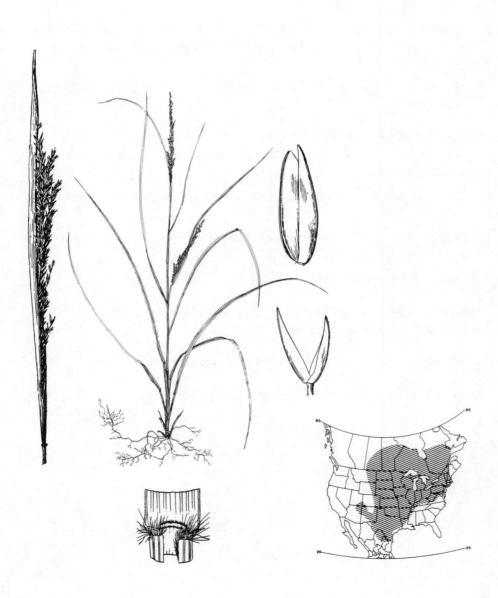

130

Tribe	ERAGROSTEAE
Species	*Sporobolus asper* (Michx.) Kunth
Common Name	Tall dropseed
Life Span	Perennial
Origin	Native
Season	Warm

INFLORESCENCE CHARACTERISTICS

type   panicle (5–30 cm long and 0.4–1.5 cm wide), contracted, solitary at each of the upper culm nodes

spikelets   1-flowered (3–7 mm long), densely crowded, long pointed, lemma somewhat rounded at apex, glabrous, lemma 1-nerved

glumes   keeled, unequal, first glume ½ as long as lemma, second glume ⅔ to ¾ as long as lemma, bright green midnerve

other   panicle entirely or partially enclosed in inflated sheath

VEGETATIVE CHARACTERISTICS

culm   erect (0.6–1.2 m tall), slender, stout, solitary or small tufts

sheath   oval, glabrous or lower ones pilose near the collar, split

blade   elongate (10–60 cm long and 1–4 mm wide), flat, wide at base tapering to a fine point, few long hairs on margin

ligule   membranous with short hairs (0.5 mm long), truncate

other   occasionally with short rhizomes

GROWTH CHARACTERISTICS   starts growth in late spring, flowers in August, reproduces by seeds, tillers, and rhizomes

FORAGE VALUE   fair for livestock and poor for wildlife, most palatable in spring

HABITAT   occurs on clayey to silty soils of prairies and sandy meadows

131

Sand dropseed (*Sporobolus cryptandrus*)

Tribe	ERAGROSTEAE
Species	*Sporobolus cryptandrus* (Torr.) Gray
Common Name	Sand dropseed (zacatón)
Life Span	Perennial
Origin	Native
Season	Warm

INFLORESCENCE CHARACTERISTICS

type   panicle (15–40 cm long and 2–15 cm wide), contracted, branches distant, occasionally hairy at panicle axis, inflorescence totally or partially enclosed in sheath

spikelets   1-flowered (1.5 to 2.8 mm long), densely crowded on upper part of panicle branches, overlapping, short pediceled, florets blunt

glumes   thin, pointed, unequal, first glume ½ as long as second, second glume equaling or slightly shorter than lemma

VEGETATIVE CHARACTERISTICS

culm   erect (0.3–1.2 m tall), spreading

sheath   rounded, densely hairy at throat (2–4 mm long), longer than internode

blade   flat, involute on drying (4–35 cm long and 2–8 mm wide), tapering to long and slender tip, margins slightly scabrous

ligule   membranous (0.5–3.0 mm long), ciliate, rounded

other   blade beneath inflorescence at right angle to the culm

GROWTH CHARACTERISTICS   reproduces from small seeds and tillers, starts growth in early spring, seeds mature June–August

FORAGE VALUE   fair to good for livestock and poor for wildlife

HABITAT   most commonly on sandy soils, but occurs on rocky and silty soils

Oniongrass (*Melica bulbosa*)

Tribe	MELICEAE
Species	*Melica bulbosa* Geyer *ex* Port. & Coult.
Common Name	Oniongrass (bulbous oniongrass, cebollin)
Life Span	Perennial
Origin	Native
Season	Cool

INFLORESCENCE CHARACTERISTICS

type   panicle, narrow, short and stiff spreading branches, densely flowered

spikelets   2- to 9-flowered, usually 3-flowered (0.7–2.4 cm long), overlapping, upper florets sterile

glumes   papery, blunt, purple-tipped, ⅔ to ¾ as long as spikelet, acute to obtuse

VEGETATIVE CHARACTERISTICS

culm   tufted (30–60 cm tall)

sheath   closed; glabrous, scabrous, or pubescent

blade   flat to involute (2–5 mm wide); glabrous, scabrous, or pubescent

ligule   membranous (2–5 mm long)

other   base of culm swollen into bulb

GROWTH CHARACTERISTICS   starts growth in early spring, flowers in late spring or early summer, low seed viability

FORAGE VALUE   excellent for cattle, sheep, horses, elk, and deer

HABITAT   occurs on rich sandy or clay loams of meadows, rocky woods, and hills

Buffelgrass (*Cenchrus ciliaris*)

Tribe	PANICEAE
Species	*Cenchrus ciliaris* L.
Common Name	Buffelgrass (zacate buffel)
Life Span	Perennial
Origin	Introduced (from India and Africa)
Season	Warm

INFLORESCENCE CHARACTERISTICS

type   panicle (2–13 cm long and 1–2 cm wide), dense, cylindrical

spikelets   in groups of 2–4 (2.2–5.6 mm long), subtended and enclosed by numerous bristles, bur with minute and pilose peduncle

bristles   bristles united at base (0.4–1.0 cm long), long-ciliate on inner margins, purple, flexuous, terete, connate only at base or slightly above

glumes   unequal

VEGETATIVE CHARACTERISTICS

culm   erect or geniculate-spreading (0.5–1.0 m tall), branched and knotty at the base

sheath   laterally compressed and keeled, glabrous to sparsely pilose

blade   thin and usually flat (8–30 cm long and 2.5–8.0 mm wide), scabrous or slightly pilose

ligule   membranous (1.0–1.5 mm long), ciliate

GROWTH CHARACTERISTICS   reproduces by seeds and rhizomes, not frost tolerant

FORAGE VALUE   good for both livestock and wildlife

HABITAT   sandy soils, old fields, disturbed sites, and seeded in pastures

Arizona cottontop (*Digitaria californica*)

Tribe	PANICEAE
Species	*Digitaria californica* (Benth.) Henr.
Common Name	Arizona cottontop (zacate punta blanca)
Life Span	Perennial
Origin	Native
Season	Warm

INFLORESCENCE CHARACTERISTICS

type   panicle (8–15 cm long) erect, narrow, alternate racemose branches (3–5 cm long), ascending

spikelets   in pairs, covered and exceeded by white to purple silky hairs; lemma awn-tipped or mucronate

glumes   second glume densely villous

VEGETATIVE CHARACTERISTICS

culm   erect (0.4–1.0 m tall); base knotty, swollen, and pubescent; nodes pubescent; often purple

sheath   lower sheaths felty pubescent or sparsely hairy

blade   narrow (2–12 cm long and 2–7 mm wide), pubescent, occasional glandular hairs on lower surface, margin toothed, midrib prominent

ligule   membranous (2–3 mm long), rounded, erose

GROWTH CHARACTERISTICS   starts growth in late spring or early summer when moisture becomes available, reproduces by seed, and seed set is good

FORAGE VALUE   good for livestock and fair for wildlife, palatable throughout the year, cures well

HABITAT   plains, dry open ground, chaparral, and semidesert grass-land

Maidencane (*Panicum hemitomon*)

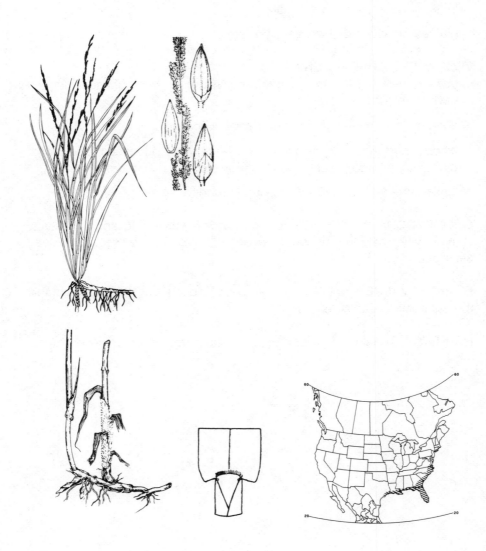

Tribe	PANICEAE
Species	*Panicum hemitomon* Schult.
Common Name	Maidencane
Life Span	Perennial
Origin	Native
Season	Warm

INFLORESCENCE CHARACTERISTICS

type   panicle (10–30 cm long and seldom over 1.5 cm wide), ascending primary branches and appressed secondary branches

spikelets   crowded, subsessile (2.0–2.7 mm long), lanceolate, awnless

glumes   first glume ½ the length of the spikelet, scabrous on midnerve, somewhat beaked at apex

VEGETATIVE CHARACTERISTICS

culm   erect (0.5–1.5 m tall), few floriferous culms

sheath   usually glabrous on flowering culms, densely hirsute on sterile culms

blade   lanceolate, flat (12–30 cm long and 0.7–1.5 cm wide) glabrous to scabrous on upper surface

ligule   densely ciliate membrane (1 mm long including hairs)

other   scaly, creeping rhizomes; often producing numerous sterile shoots with overlapping sheaths

GROWTH CHARACTERISTICS   new shoots arise from rhizomes from January–March, seed is set during June and July, and a second growth period may occur from October–December

FORAGE VALUE   excellent for livestock and fair for wildlife

HABITAT   moist soil along river banks and ditches, borders of lakes and ponds, often in the water, and sometimes a weed in cultivated fields

Vinemesquite (*Panicum obtusum*)

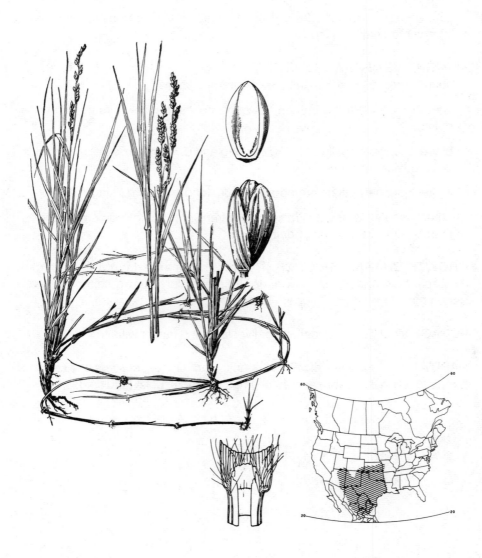

Tribe	PANICEAE
Species	*Panicum obtusum* H.B.K.
Common Name	Vinemesquite (zacate guia)
Life Span	Perennial
Origin	Native
Season	Warm

INFLORESCENCE CHARACTERISTICS

type   panicle (3–14 cm long and 0.5–1.3 cm wide), narrow, primary branches usually unbranched, panicle branches distant, erect

spikelets   2-flowered, oblong or obovate (3–4 mm long), glabrous, hard, awnless

glumes   first glume as long as spikelet, brown at maturity

VEGETATIVE CHARACTERISTICS

culm   erect (20–80 cm tall), tufted from a knotty base, nodes hairy

sheath   rounded (½–¾ the length of the internode), glandular hairs, collar often hairy, sometimes sparingly pilose near margins

blade   firm, elongate (5–20 cm long and 2–7 mm wide), white midrib prominent abaxially

ligule   membranous (1–2 mm long), erose

other   long wiry stolons from a hard, knotty, or rhizomatous base

GROWTH CHARACTERISTICS   starts growth in April–May; reproduces by seeds, rhizomes, and stolons; may grow in pure stands

FORAGE VALUE   good for livestock and fair for wildlife, withstands heavy grazing

HABITAT   sandy or gravely soil; along banks of rivers, irrigation ditches, and clayey lowland pastures

Switchgrass (*Panicum virgatum*)

Notes:

Tribe	PANICEAE
Species	*Panicum virgatum* L.
Common Name	Switchgrass
Life Span	Perennial
Origin	Native
Season	Warm

INFLORESCENCE CHARACTERISTICS

type panicle (15–55 cm long), diffuse, spikelets toward ends of long branches, branches in whorls

spikelets 2-flowered with upper floret fertile (3–5 mm long), glabrous, margins of lemma inrolled at base, awnless; lemma of fertile floret is smooth and shiny; lower floret sterile or staminate

glumes unequal, narrowly acute or acuminate, first glume ¾ as long as spikelet and encircles the base at the second glume

VEGETATIVE CHARACTERISTICS

culm erect (0.5–3.0 m tall), robust, usually unbranched above base

sheath rounded, often purple to red at base

blade firm, flat, elongate (10–60 cm long and 0.3–1.5 cm wide), adaxial side at base with triangular patch of hair, margin weakly barbed

ligule fringed membrane (1.5–3.5 mm long), mostly hairs, rounded

other rhizomatous

GROWTH CHARACTERISTICS reproduces from seeds, tillers, and rhizomes; starts growth in April, and may grow in clumps

FORAGE VALUE good for all types of livestock and fair for wildlife; as the stems begin to mature in midsummer, nutrient content and palatability decline rapidly

HABITAT prairies, open ground, open woods, brackish marshes, and pine woods

Dallisgrass (*Paspalum dilatatum*)

Tribe	PANICEAE
Species	*Paspalum dilatatum* Poir.
Common Name	Dallisgrass
Life Span	Perennial
Origin	Introduced (from South America)
Season	Warm

INFLORESCENCE CHARACTERISTICS

type   panicle of 2–5 straight racemose branches (3–8 cm long), widely spaced on a slender axis

spikelets   solitary, in 4 rows on 1 side of branch, rounded, fringed with silky hair, awnless

glumes   first glume absent, second glume 3- to 5-nerved (3–4 mm long)

VEGETATIVE CHARACTERISTICS

culm   erect (0.5–1.5 m tall), tufted, knotty, and decumbent base

sheath   lower sheaths pubescent, upper sheaths glabrous

blade   flat, firm, tapering to a point (10–25 cm long and 0.3–1.2 cm wide), glabrous or sparsely ciliate with long hairs near base

ligule   membranous (1.5–3.0 mm long), rounded, entire, brown

GROWTH CHARACTERISTICS   reproduces by seed and rhizomes, starts growth in March–April, drought tolerant

FORAGE VALUE   good for livestock and fair for wildlife

HABITAT   low ground, dry prairies, marshy meadows, waste places, seeded in pastures

Plains bristlegrass (*Setaria leucopila*)

Tribe	PANICEAE
Species	*Setaria leucopila* (Scribn. & Merr.) K. Schum.
Common Name	Plains bristlegrass (zacate tempranero)
Life Span	Perennial
Origin	Native
Season	Warm

INFLORESCENCE CHARACTERISTICS

type   panicle, cylindrical (6–25 cm long and 0.6–1.5 cm wide), densely flowered

spikelets   small (2–3 mm long), subtended by single bristle (0.4–1.5 cm long), fruit rugose, awnless

glumes   unequal, first glume reduced, second as long as spikelet

VEGETATIVE CHARACTERISTICS

culm   stiffly erect or geniculate (0.2–1.2 m tall), tufted, often hairy below the nodes, infrequently branched above

sheath   often villous on upper margins, keeled

blade   flat or folded (8–40 cm long and 2–8 mm wide); glabrous to scabrous or infrequently pubescent

ligule   membranous (1–2 mm long), fringed with hairs, rounded

GROWTH CHARACTERISTICS   starts growth mid-spring; may produce more than one seed crop depending on available moisture; good seed producer

FORAGE VALUE   good for livestock and fair for wildlife

HABITAT   open dry ground and dry woods

Ripgut brome (*Bromus diandrus*)

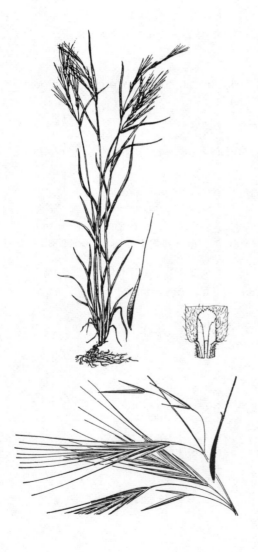

150

Tribe	POEAE
Species	*Bromus diandrus* Roth
Common Name	Ripgut brome
Life Span	Annual
Origin	Introduced (from Europe)
Season	Cool

INFLORESCENCE CHARACTERISTICS

type   panicle (7–15 cm long), narrow, stout and erect branches, pubescent

spikelets   large (3–4 cm long excluding awns), 4- to 8-flowered, lemmas glabrous or scabrous with broad hyaline margins and long teeth (4–5 mm long)

awns   lemmas awned, stout (3–6 cm long)

glumes   unequal (1.5–3.0 mm long), smooth, margins hyaline, barbs on back

VEGETATIVE CHARACTERISTICS

culm   thick, weak (20–70 cm tall)

sheath   pilose, spreading

blade   flat, soft (10–15 cm long and 4–7 mm wide), pilose

ligule   membranous (3–5 mm long), erose-lacerate

GROWTH CHARACTERISTICS
seeds germinate in late fall, matures 2–3 months after beginning rapid spring growth

FORAGE VALUE
excellent in seedling stage and during vigorous vegetative growth, but becomes poor for sheep and wildlife and fair for cattle at flowering

HABITAT
common weed in open ground and waste places

Smooth brome (*Bromus inermis*)

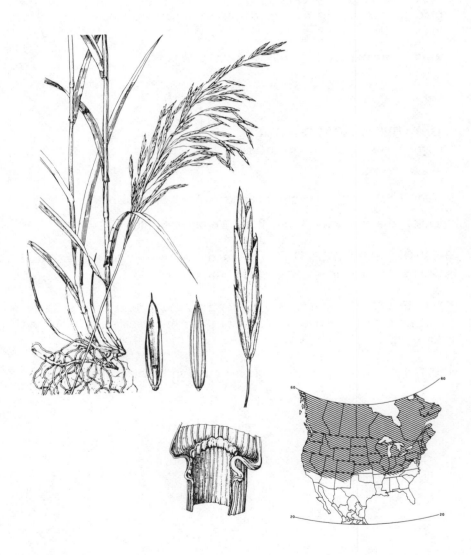

Tribe	POEAE
Species	*Bromus inermis* Leyss.
Common Name	Smooth brome (bromo suave)
Life Span	Perennial
Origin	Introduced (from Europe)
Season	Cool

INFLORESCENCE CHARACTERISTICS

type   panicle, erect (7–20 cm long), branches whorled, contracted at maturity, becoming purplish or brownish

spikelets   terete, pointed, 5- to 13-flowered (1.5–3.0 cm long), lemmas rounded on back, shallow bifid apex, 3- to 5-nerved

awns   lemmas awnless or with a short awn (1–2 mm)

glumes   papery, lanceolate, unequal, first glume 1-nerved (4–6 mm long); second glume 3-nerved (0.6–1.0 cm long)

VEGETATIVE CHARACTERISTICS

culm   erect (0.5–1.0 m tall)

sheath   closed, glabrous to scabrous (occasionally hirsute-pilose), prominently veined

blade   flat (15–40 cm long and 0.4–1.5 cm wide), keeled below, glabrous to pubescent, margins scabrous, conspicuous "W" on blade

ligule   membranous (0.5–2.5 mm long), minutely ciliate-erose

other   creeping rhizomes

GROWTH CHARACTERISTICS   starts growth in early spring; reproduces by seeds, tillers, and rhizomes; responds to fertilization

FORAGE VALUE   excellent for livestock and wildlife, remaining palatable even after inflorescence development

HABITAT   cultivated as a hay and pasture grass, roadsides, waste places

Mountain brome (*Bromus marginatus*)

Tribe	POEAE
Species	*Bromus marginatus* Nees
Common Name	Mountain brome
Life Span	Perennial
Origin	Native
Season	Cool

INFLORESCENCE CHARACTERISTICS

type panicle, erect (10–25 cm long) narrow

spikelets flattened, 5- to 11-flowered (2.5–4.0 cm long), short branches

awns lemmas awned (5–7 mm)

glumes unequal, first glume shorter and 3- to 5-nerved, second glume 5- to 7-nerved

VEGETATIVE CHARACTERISTICS

culm erect (0.3–1.2 m tall), tufted, coarse, may develop rhizomes

sheath retrorsely to antrorsely pilose or nearly glabrous, closed to near the throat

blade flat (15–40 cm long and 0.6–1.0 cm wide), pubescent abaxially

ligule membranous (1.5–3.0 mm long), truncate-acute, usually erose

GROWTH CHARACTERISTICS starts growth in early spring, seeds mature by August, reproduces from seeds and tillers

FORAGE VALUE excellent for livestock and wildlife, becoming harsh and fibrous at maturity

HABITAT open woods, open or wooded slopes, meadows, and waste places

Soft brome (*Bromus mollis*)

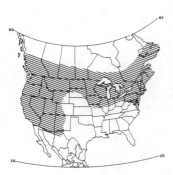

Tribe	POEAE
Species	*Bromus mollis* L.
Common Name	Soft brome (soft chess)
Life Span	Annual
Origin	Introduced (from Europe)
Season	Cool

INFLORESCENCE CHARACTERISTICS

type   panicle (3–10 cm long), erect, compact

spikelets   5- to 9-flowered (1.5–2.0 cm long); lemmas broad, soft, obtuse, pilose or scabrous, deeply toothed, margin hyaline

awns   lemma awned (5–9 mm long)

glumes   large, broad, obtuse, pilose or scabrous, first glume 3- to 5-nerved, second glume 5- to 7-nerved

other   pedicels and branches typically shorter than spikelets

VEGETATIVE CHARACTERISTICS

culm   weak, geniculate (20–80 cm tall), softly pubescent

sheath   closed, densely hirsute with spreading hairs

blade   soft, flat, or folded (3–15 cm long and 2–6 mm wide); glabrous or sparsely hirsute

ligule   membranous (0.5–1.0 mm long), erose, truncate to rounded

GROWTH CHARACTERISTICS   germinates in late fall when moisture is adequate, reproduces from seeds, growth period of about 12 weeks

FORAGE VALUE   excellent for livestock and good for wildlife while immature, good to fair when mature, soft awns allow it to be grazed without injury even after seed maturity

HABITAT   open ground waste places, well drained soils

Downy brome (Bromus tectorum)

158

Tribe	POEAE
Species	*Bromus tectorum* L.
Common Name	Downy brome (cheatgrass, broncograss)
Life Span	Annual
Origin	Introduced (from Europe)
Season	Cool

INFLORESCENCE CHARACTERISTICS

type   panicle (5–15 cm long), dense, flexuous, drooping, branches and pedicels slender, often purple

spikelets   5- to 8-flowered (1.2–2.0 cm long excluding awns); lemmas (0.9–1.2 cm long) with thin membranous margins and slender apical teeth (2–3 mm long)

awns   lemma awned (1.2–1.8 cm long)

glumes   unequal, first glume 4–6 mm long, second glume 0.8–1.0 cm long, broad hyaline margins, villous

VEGETATIVE CHARACTERISTICS

culm   erect or spreading (25–60 cm tall), weak

sheath   round, keeled toward collar, softly pubescent

blade   pubescent (5–12 cm long and 3–7 mm wide)

ligule   membranous (2–3 mm long), acute, erose-lacerate

GROWTH CHARACTERISTICS   seeds germinate in the late fall or early spring, rapid spring growth, seeds mature about 2 months later

LIVESTOCK LOSSES   after maturity awns may cause eye injury or jaw abscesses

FORAGE VALUE   fair to good for livestock and wildlife before the inflorescence emerges, then is practically worthless

HABITAT   heavily grazed rangeland, roadsides, waste places, disturbed sites

Orchardgrass (*Dactylis glomerata*)

Notes:

Tribe	POEAE
Species	*Dactylis glomerata* L.
Common Name	Orchardgrass (cocksfoot, zacate orchard)
Life Span	Perennial
Origin	Introduced (from Europe and Asia)
Season	Cool

INFLORESCENCE CHARACTERISTICS

type   panicle (3–20 cm long), branches distant, branches bare of spikelets on lower ½, branches stiffly ascending except some lower ones that spread

spikelets   tightly clustered on one side of branch, nearly sessile, 5-flowered (5–9 mm long)

awns   glumes and lemmas mucronate to awn-tipped (up to 1 mm long)

glumes   unequal, keeled, pubescent on keel

VEGETATIVE CHARACTERISTICS

culm   erect (0.5–1.2 m tall), large clumps, glabrous

sheath   flat, often keeled and laterally compressed, glabrous to slightly scabrous

blade   elongate (10–40 cm long and 0.2–1.1 cm wide), flat or folded, prominent and scabrous midrib, scabrous margins

ligule   membranous (2–8 mm long), entire, rounded

other   rarely with short rhizomes

GROWTH CHARACTERISTICS   shade tolerant, reproduces by seeds and tillers, starts growth in early spring

FORAGE VALUE   good to excellent for livestock and wildlife, provides early spring forage

HABITAT   grows on fine or coarse soils of fields, meadows, and waste places; commonly seeded in pastures

Idaho fescue (*Festuca idahoensis*)

Notes:

Tribe	POEAE
Species	*Festuca idahoensis* Elmer
Common Name	Idaho fescue
Life Span	Perennial
Origin	Native
Season	Cool

INFLORESCENCE CHARACTERISTICS

type   panicle (10−20 cm long), narrow, loose, lower branches spreading

spikelets   5- to 7-flowered (1.0−1.4 cm long), lemma cylindrical to some-what flattened (7 mm long)

awns   lemma awned (2−4 mm long)

glumes   unequal, shorter than lemma, acute

VEGETATIVE CHARACTERISTICS

culm   erect (0.3−1.0 m tall), densely tufted

sheath   flattened, keeled, glabrous or scabrous, basal sheaths short

blade   elongate (5−25 cm long), involute, filiform, very scabrous, often glaucous

ligule   membranous (less than 2 mm long), ciliate

GROWTH CHARACTERISTICS   reproduces by seed, starts growth in early spring, seeds mature by midsummer

FORAGE VALUE   excellent for livestock and wildlife, especially important late in the growing season

HABITAT   grows on all exposures and on many soil types on rocky slopes and in open woods

Rough fescue (*Festuca scabrella*)

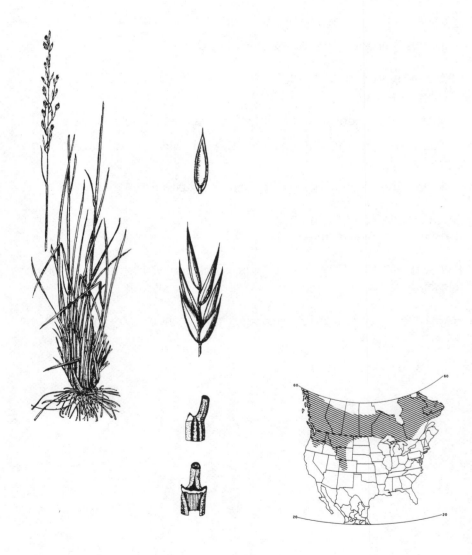

Tribe	POEAE
Species	*Festuca scabrella* Torr.
Common Name	Rough fescue
Life Span	Perennial
Origin	Native
Season	Cool

INFLORESCENCE CHARACTERISTICS

type   panicle (5–25 cm long), narrow, branches usually in pairs

spikelets   4- to 6-flowered, often purple; lemmas stout (0.7–1.0 cm long), scabrous

awns   lemma acute (awnless) or short-awned (2–4 mm long)

glumes   unequal, first glume lanceolate, 1-nerved and slightly shorter than second glume, second glume 3-nerved

VEGETATIVE CHARACTERISTICS

culm   erect (30–90 cm tall), densely tufted, large bunches (up to 30 cm in diameter), scabrous below panicle, stout

sheath   enlarged at base; margins broad, hyaline, and smooth to scabrous

blade   folded becoming involute, (30–70 cm long and 0.8–1.6 cm wide), sharp-pointed, usually scabrous

ligule   membranous (less than 1 mm long), ciliate, truncate

other   purple at base

GROWTH CHARACTERISTICS   growth period is June–July, reproduces from seeds, tillers, and occasionally has rhizomes

FORAGE VALUE   excellent for livestock and good for wildlife during all growth stages

HABITAT   mountains and foothills

Mutton bluegrass (*Poa fendleriana*)

Notes:

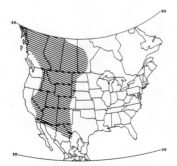

Tribe	POEAE
Species	*Poa fendleriana* (Steud.) Vasey
Common Name	Mutton bluegrass (muttongrass)
Life Span	Perennial
Origin	Native
Season	Cool

INFLORESCENCE CHARACTERISTICS

type   panicle (2–10 cm long), narrow, densely flowered, 2–3 branches per node, branches erect or erect-spreading

spikelets   3- to 8-flowered (0.6–1.0 cm long), flat, usually twice as long as wide; florets papery; lemma marginal nerves hairy

glumes   broad, thin, papery, usually ½ to ⅔ as long as lowest lemma, subequal, strongly keeled

VEGETATIVE CHARACTERISTICS

culm   erect (15 to 80 cm tall), tufted, rough below inflorescence, decumbent base

sheath   short, margin hyaline, sheath bases white and expanded

blade   stiff, folded or involute (10–20 cm long and 1–4 mm wide), leaves short basal, scabrous, glaucous, often remain green

ligule   membranous (about 0.5 mm long, occasionally up to 3 mm long), obtuse or truncate

other   seldom has slender rhizomes

GROWTH CHARACTERISTICS   starts growth in early spring and matures by June or July, reproduces by seeds, tillers, and rarely by rhizomes

FORAGE VALUE   excellent for cattle and horses, good for sheep and wildlife

HABITAT   mesas, open dry woods, rocky hills

Kentucky bluegrass (*Poa pratensis*)

Notes:

Tribe	POEAE
Species	*Poa pratensis* L.
Common Name	Kentucky bluegrass (zacate azul de Kentucky)
Life Span	Perennial
Origin	Introduced (from Europe)
Season	Cool

INFLORESCENCE CHARACTERISTICS

type   panicle (3–13 cm long and 3–8 cm wide), pyramidal, open long panicle branches, lower branches in whorl of 3–5 (commonly 5)

spikelets   3- to 6-flowered (3–6 mm long), flattened, nearly as wide as long; lemmas with a tuft of long, silky hairs at base

glumes   slightly unequal (2.0–3.5 mm long), strongly keeled, scabrous on keel

VEGETATIVE CHARACTERISTICS

culm   slender (0.2–1.0 m tall), tufted wiry, curving-erect from creeping rhizomatous base

sheath   glabrous or scabrous, keeled, distinctly veined, closed about ½ of the length

blade   folded or flat (5–40 cm long and 1–5 mm wide), elongated, keeled

ligule   membranous (1–2 mm long), truncate, entire

GROWTH CHARACTERISTICS   rhizomes initiated in summer or fall; initiates aerial culms in spring and summer; reproduces by seeds, tillers, and rhizomes

FORAGE VALUE   good for livestock and wildlife

HABITAT   meadows, open woods, open ground, weed on disturbed sites, commonly planted for lawns and some pastures

Sandberg bluegrass (*Poa sandbergii*)

Tribe	POEAE
Species	*Poa sandbergii* Vasey
Common Name	Sandberg bluegrass
Life Span	Perennial
Origin	Native
Season	Cool

INFLORESCENCE CHARACTERISTICS

type   panicle (2–10 cm long), narrow, 2 to 3 branches per node; branches unequal in length, appressed, yellowish

spikelets   2- to 5-flowered (4–6 mm long), terete, longer than wide; lemma short-pubescent on back of base, convex on back, purple

glumes   unequal, papery, second glume (3–4 mm long) is shorter than lemma of lowermost floret

VEGETATIVE CHARACTERISTICS

culm   erect (10–45 cm tall), tufted, glabrous, nodes occasionally red

sheath   glabrous, veins prominent, margins hyaline and overlapping

blade   basal, short (3–16 cm long and 1–3 mm wide); flat, folded or involute; glabrous, double midrib, margin slightly barbed

ligule   membranous (2–4 mm long), acute

GROWTH CHARACTERISTICS   starts growth in early spring, seeds mature in early summer, reproduces by seeds and tillers; drought enduring

FORAGE VALUE   good for cattle and fair for sheep and wildlife in spring and early summer

HABITAT   usually on shallow soils of plains, dry woods, and rocky slopes

Sixweeksgrass (*Vulpia octoflora*)

Tribe	POEAE
Species	*Vulpia octoflora* (Walt.) Rydb.
Common Name	Sixweeksgrass (sixweeks fescue)
Life Span	Annual
Origin	Native
Season	Cool

INFLORESCENCE CHARACTERISTICS

type   panicle (1–20 cm long), narrow, short appressed branches

spikelets   glabrous, scabrous, or pubescent; 5- to 17-flowered (0.4–1.0 cm long excluding the awns), dense, herringbone pattern, translucent

awns   lemma awned (3–7 mm long)

glumes   unequal, subulate-lanceolate

VEGETATIVE CHARACTERISTICS

culm   erect (occasionally decumbent), weak (10–60 cm tall), solitary or loosely tufted

sheath   keeled and ridged, sparingly pubescent

blade   narrow (2–10 cm long and 0.5–2.0 mm wide), involute, sparingly pubescent, margins slightly ciliate

ligule   membranous (0.5–1.0 mm long), truncate, erose

GROWTH CHARACTERISTICS   starts growth in early spring and may mature 6 weeks later, reproduces by seed

FORAGE VALUE   little forage value for livestock or wildlife except during a brief period in early spring

HABITAT   open ground, disturbed areas

Indian ricegrass (*Oryzopsis hymenoides*)

174

Tribe	STIPEAE
Species	*Oryzopsis hymenoides* (R. & S.) Ricker
Common Name	Indian ricegrass
Life Span	Perennial
Origin	Native
Season	Cool

INFLORESCENCE CHARACTERISTICS

type   panicle (7−25 cm long), diffuse, dichotomous branching

spikelets   1-flowered (5−8 mm long excluding awn), on curved pedicels (0.5−3.0 cm long), solitary; lemma pubescent on callus (hairs 2−4 mm long), lemma brown to black at maturity

awns   lemma awned (3−6 mm), stout, straight, early deciduous; glumes sometimes short-awned

glumes   subequal (5−8 mm long), broad, thin, glabrous to puberulent

VEGETATIVE CHARACTERISTICS

culm   stiffly erect (30−70 cm tall), densely tufted, glabrous

sheath   glabrous, rounded on the back, shorter than the internodes, ciliate on one margin, collar pubescent

blade   long (5−30 cm long and 1−2 mm wide), involute, filiform, midrib prominent abaxially

ligule   membranous (3−7 mm long), acuminate, may be deeply notched

GROWTH CHARACTERISTICS   starts growth in early spring, reproduces by seeds; one of the drought enduring species

FORAGE VALUE   good for livestock and wildlife, especially valuable for winter grazing, seeds high in protein, used to make flour by various Indian tribes

HABITAT   plains and deserts; dry sandy and silty soils and disturbed sites

Smilo (*Oryzopsis miliacea*)

176

Tribe	STIPEAE
Species	*Oryzopsis miliacea* (L.) Asch. & Schweinf.
Common Name	Smilo
Life Span	Perennial
Origin	Introduced (from the Mediterranean region)
Season	Cool

INFLORESCENCE CHARACTERISTICS

type panicle (15–30 cm long), loose, branches spreading, lower nodes with many long branches, lower branches in whorls

spikelets 1-flowered, short-pediceled, lemma smooth (1.5–2.5 mm long)

awns lemma awned (4 mm long), deciduous

glumes equal (3 mm long), acuminate

VEGETATIVE CHARACTERISTICS

culm erect (0.6–1.5 m tall), stout, from decumbent stem

sheath flat, glabrous

blade flat (20–50 cm long and 0.5–1.0 cm wide)

ligule membranous (2–5 mm long), acute

other leaves may be brown when mature while culms remain green

GROWTH CHARACTERISTICS
stout perennial, starts growth in early spring when moisture is available, reproduces by seeds and tillers

FORAGE VALUE
good for livestock and fair for wildlife when the plants are immature

HABITAT
dry soils, mostly seeded for forage on converted brushlands

Columbia needlegrass (*Stipa columbiana*)

178

Tribe	STIPEAE
Species	*Stipa columbiana* Macoun
Common Name	Columbia needlegrass (subalpine needlegrass)
Life Span	Perennial
Origin	Native
Season	Cool

INFLORESCENCE CHARACTERISTICS

type   panicle (7–30 cm long), narrow, dense, branches appressed, often purple

spikelets   1-flowered, sharp-pointed callus; lemma slender (5–7 mm long), pubescent, apex hairs longest (0.7–1.0 mm long)

awns   lemma awned (1.8–2.5 cm long), twice-geniculate, minutely scabrous and twisted below, upper segment often 1.5 cm or longer

glumes   subequal, exceed lemma, 3- or 5-nerved

VEGETATIVE CHARACTERISTICS

culm   erect (0.3–1.0 m tall), tufted, straight, nodes may be purple

sheath   naked at throat, round and slightly keeled, glabrous to scabrous, occasionally pubescent on margins at apex, open

blade   basal, involute (10–30 cm long and 1–4 mm wide), margin barbed, glabrous, narrower at base than sheath apex

ligule   membranous (0.5–2.0 mm long), erose, truncate

GROWTH CHARACTERISTICS   starts growth in mid-spring and matures by September, reproduces by seeds and tillers

LIVESTOCK LOSSES   pointed callus may work into the ears and tongue

FORAGE VALUE   fair to good for cattle, and sheep, and fair for wildlife

HABITAT   dry plains, meadows, and open woods in foothills and mountains

Needleandthread (*Stipa comata*)

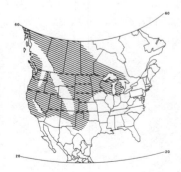

180

Tribe	STIPEAE
Species	*Stipa comata* Trin. & Rupr.
Common Name	Needleandthread
Life Span	Perennial
Origin	Native
Season	Cool

INFLORESCENCE CHARACTERISTICS

type   panicle (10–20 cm long), contracted, remains partially in sheath

spikelets   1-flowered, large; lemma large (1.0–1.5 cm long), pale to brown at maturity, callus bearded with stiff hairs

awns   lemma awned (10–20 cm long), loosely twisted, flexuous, twisted and short-pubescent on lower ½, terminal segment glabrous or scabrous

glumes   about equal (1.5–3.5 cm long), narrow, attenuate tips sub-hyaline

VEGETATIVE CHARACTERISTICS

culm   erect (0.3–1.1 m tall), densely tufted, glabrous

sheath   round, open, glabrous or scabrous, prominently veined

blade   flat or involute (5–40 cm long and 1–3 mm wide)

ligule   membranous (2–4 mm long), split or widely notched at top

GROWTH CHARACTERISTICS   starts growth in early spring or when moisture is available and reproduces by seed

LIVESTOCK LOSSES   sharp-pointed callus and long awns may cause injury by working into the eyes, tongue, and ears; sheep are especially susceptible to injury

FORAGE VALUE   fair to good for livestock and poor to fair for wildlife, cures well to provide fall and winter forage

HABITAT   prairies, plains, and dry hills

Texas wintergrass (*Stipa leucotricha*)

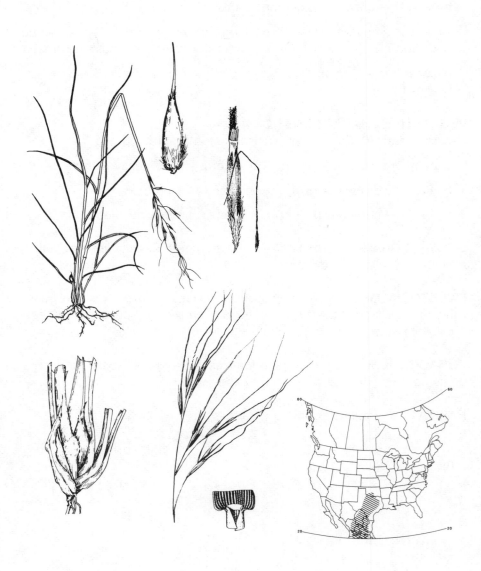

Tribe	STIPEAE
Species	*Stipa leucotricha* Trin. & Rupr.
Common Name	Texas wintergrass (Texas needlegrass)
Life Span	Perennial
Origin	Native
Season	Cool

INFLORESCENCE CHARACTERISTICS

type   panicle (6–25 cm long), narrow (usually less than 10 cm wide); slender, flexuous lower branches

spikelets   1-flowered; lemma oblong (0.9–1.2 cm long), light brown, appressed pubescent below, rugose on body above base, rounded and white neck with ring of stiff hairs (0.5–1.0 mm long)

awns   lemma awned (4–10 cm long), stout, once- or twice-geniculate, scabrous-pubescent on twisted lower portion (2.0–3.5 cm)

glumes   about equal (1.2–1.8 cm long), thin, glabrous

VEGETATIVE CHARACTERISTICS

culm   tufted (30–90 cm tall), spreading at base, nodes pubescent

sheath   pubescent or nearly glabrous; collar with long hairs on sides

blade   flat, becoming involute (5–40 cm long and 1–5 mm wide), usually pubescent with short and stiff hairs on one or both surfaces

ligule   variable, membranous (absent to 1 mm long), truncate

GROWTH CHARACTERISTICS   starts growth in late fall, remains green through winter and spring, and reproduces from seeds and cleistogamous spikelets

LIVESTOCK LOSSES   awns may cause some injury

FORAGE VALUE   fair for both livestock and wildlife, considerable value for early spring forage

HABITAT   dry grasslands, disturbed sites, and heavily grazed pastures

Purple needlegrass (*Stipa pulchra*)

Tribe	STIPEAE
Species	*Stipa pulchra* Hitchc.
Common Name	Purple needlegrass
Life Span	Perennial
Origin	Native
Season	Cool

INFLORESCENCE CHARACTERISTICS

type panicle (15−20 cm long), narrow, loose, nodding with spreading branches, lower branches 2.5−5.0 cm long

spikelets 1-flowered; lemma narrow (0.7−1.3 cm long), fusiform, sparingly pilose, summit sometimes with smooth neck and ciliate crown

awns lemma awned (7−9 cm long), twice-geniculate, short-pubescent to second bend; first segment 1.5−2.0 cm long, second shorter, third 4−6 cm long

glumes unequal, narrow, long acuminate, purple, 3-nerved

VEGETATIVE CHARACTERISTICS

culm erect (0.6−1.0 m tall), tufted, pubescent only directly below the nodes

sheath open

blade flat or involute (10−30 cm long and 3−5 mm wide), basal leaves numerous

ligule membranous (1 mm long)

GROWTH CHARACTERISTICS starts growth in late fall or early spring, flowers April−June, reproduces by seeds and tillers

LIVESTOCK LOSSES awns may cause livestock injury similar to *Stipa comata*

FORAGE VALUE good for livestock and fair for wildlife

HABITAT prairies, open ground, waste places, disturbed sites

Thurber needlegrass (*Stipa thurberiana*)

Tribe	STIPEAE
Species	*Stipa thurberiana* Piper
Common Name	Thurber needlegrass
Life Span	Perennial
Origin	Native
Season	Cool

INFLORESCENCE CHARACTERISTICS

type   panicle (8–15 cm long), narrow, branches few-flowered, purple

spikelets   1-flowered, lemma (8–9 mm long) with pointed callus

awns   lemma awned (4–5 cm long), twice-geniculate, plumose on the first and second segments (hairs 1–2 mm long)

glumes   nearly equal (1.1–1.3 cm long), acuminate

VEGETATIVE CHARACTERISTICS

culm   erect (30–60 cm tall), tufted

sheath   scabrous, upper sheath occasionally glabrous

blade   involute (10–15 cm long and 1–2 mm wide), narrow, fine, scabrous to densely pubescent, flexuous

ligule   membranous (3–6 mm long)

GROWTH CHARACTERISTICS   starts growth in early spring, seeds mature by July, this species is known to hybridize with *Oryzopsis hymenoides,* reproduces by seeds and tillers

LIVESTOCK LOSSES   sharp awns may cause injury similar to *Stipa comata*

FORAGE VALUE   good for livestock and fair for wildlife until seeds mature; following seed drop, animals may resume grazing

HABITAT   mesas and slopes

Crested wheatgrass (*Agropyron cristatum*)

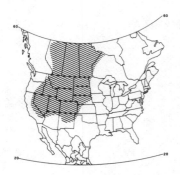

Tribe	TRITICEAE
Species	*Agropyron cristatum* (L.) Gaertn.
Common Name	Crested wheatgrass
Life Span	Perennial
Origin	Introduced (from eastern Europe and Asia)
Season	Cool

INFLORESCENCE CHARACTERISTICS

type   spike (2−9 cm long), rachis pubescent; spikelets closely imbricate, several times longer than internodes; spreading-ascending on rachis; rachis occasionally wavy

spikelets   5- to 8-flowered (0.5−1.5 cm long), much compressed, located sideways to rachis

awns   glumes and lemmas tapering to awns (2−5 mm long)

glumes   somewhat contorted, pubescent

VEGETATIVE CHARACTERISTICS

culm   erect (0.2−1.0 m tall), tufted, occasionally geniculate

sheath   glabrous (sometimes pubescent below), margins overlapping; slender auricles (1 mm long)

blade   flat (5−20 cm long and 0.2−1.0 cm wide), nerves raised above, smooth below, margins weakly scabrous

ligule   membranous (0.5−1.5 mm long), margin short-fringed to short-ciliate, rounded

GROWTH CHARACTERISTICS   starts growth in early spring, reproduces by seeds and tillers, drought resistant, grows in both spring and fall

FORAGE VALUE   good for livestock and fair for wildlife, cures well for use as winter forage

HABITAT   dry soils, planted in pastures, hay meadows, and roadsides; major reseeded species in sagebrush areas

Intermediate wheatgrass (*Agropyron intermedium*)

Notes:

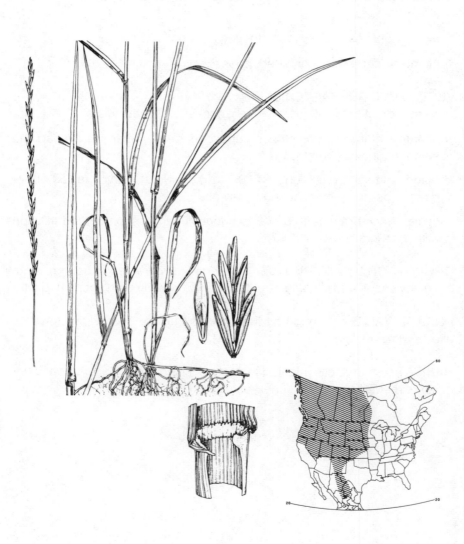

190

Tribe	TRITICEAE
Species	*Agropyron intermedium* (Host) Beauv.
Common Name	Intermediate wheatgrass
Life Span	Perennial
Origin	Introduced (from Europe and Russia)
Season	Cool

INFLORESCENCE CHARACTERISTICS

type   spike (10–20 cm long); spikelets slightly imbricate or none, may be curved away from the rachis at maturity

spikelets   4- to 8-flowered (1–2 cm long), 1 spikelet per node; lemmas glabrous (4–7 mm long), blunt

glumes   about ½ as long as the spikelet, glabrous, notched, blunt or with a slight mucronate tip, distinct nerves

VEGETATIVE CHARACTERISTICS

culm   erect (0.9–1.2 m tall), robust

sheath   open, mostly glabrous, acute auricles

blade   flat (10–40 cm long and 0.5–1.0 cm wide), broad at base and tapers to a point, strong veined

ligule   membranous (1–2 mm long), truncate

other   rhizomes

GROWTH CHARACTERISTICS   starts growth in early spring, matures June–August, reproduces by seed, tillers, and rhizomes

FORAGE VALUE   good to excellent for all classes of livestock and fair for wildlife

HABITAT   uplands, moist to dry soils, seeded in both dryland and irrigated pastures and hay meadows

Western wheatgrass (*Agropyron smithii*)

Tribe	TRITICEAE
Species	*Agropyron smithii* Rydb.
Common Name	Western wheatgrass (bluestem, bluejoint)
Life Span	Perennial
Origin	Native
Season	Cool

INFLORESCENCE CHARACTERISTICS

type   spike (6–20 cm long), often dense, spikelets closely imbricate (about ½ of each spikelet overlaps), occasionally 2 spikelets per node

spikelets   5- to 12-flowered (1.5–2.5 cm long), glaucous; lemmas several-nerved, glabrous to pubescent on margins

awns   glumes occasionally awn-tipped, lemmas occasionally awn-tipped

glumes   asymmetrical, faintly nerved, rigid, narrow

other   occasionally dense pubescence on spikelet

VEGETATIVE CHARACTERISTICS

culm   erect (30–90 cm tall), single or in small clusters, glaucous

sheath   glabrous or scabrous, with or without auricles (1–2 mm long)

blade   rigid, tapering to a sharp point (10–25 cm long and 2–6 mm wide), strongly veined, often involute on drying, glaucous

ligule   membranous (1 mm long), truncate, minutely ciliate

other   rhizomes

GROWTH CHARACTERISTICS   reproduces from seeds and rhizomes, growth starts when daytime temperatures are 12–13° C, dormat in summer, begins growth again in fall

FORAGE VALUE   good for all classes of livestock, fair for antelope and other wildlife; cures well making good winter forage

HABITAT   in a wide variety of soils on low, moist flats or floodplains; scattered on dry sites

Bluebunch wheatgrass (*Agropyron spicatum*)

Tribe	TRITICEAE
Species	*Agropyron spicatum* (Pursh) Scribn.
Common Name	Bluebunch wheatgrass
Life Span	Perennial
Origin	Native
Season	Cool

INFLORESCENCE CHARACTERISTICS

type   spike (8–15 cm long), slender, rachis internodes 1–2 cm long, spikelets not imbricate to ⅛ imbricate, 1 spikelet per node

spikelets   6- to 8-flowered (1.0–2.5 cm long); lemmas about 1 cm long

awns   lemmas awned (1–2 cm long), strongly divergent at maturity, not awned in *A. spicatum* var. *inerme*

glumes   narrow, obtuse to acute, about ½ as long as spikelet

VEGETATIVE CHARACTERISTICS

culm   erect (0.6–1.0 m tall), tufted, slender, pubescent nodes

sheath   glabrous to appressed-puberulent, margins overlapping, strongly veined; auricles acute, clasping, red, small (when present)

blade   flat to loosely involute (5–25 cm long and 1–5 mm wide), pubescent on upper surface, margins white and weakly barbed

ligule   membranous (0.5–1.0 mm long), ciliate, rounded or truncate

other   rhizomes are very rare; plant glaucous

GROWTH CHARACTERISTICS
growth begins in April, and the plant stays green well into the summer; regrowth occurs following fall rains; reproduces by seeds and rarely by rhizomes

FORAGE VALUE
excellent for cattle and horses, good for sheep, elk, and deer; cures well and makes good standing winter feed

HABITAT
plains, dry slopes, canyons, and dry open woods

Slender wheatgrass (*Agropyron trachycaulum*)

Tribe	TRITICEAE
Species	*Agropyron trachycaulum* (Link) Malte
Common Name	Slender wheatgrass
Life Span	Perennial
Origin	Native
Season	Cool

INFLORESCENCE CHARACTERISTICS

type   spike (10–25 cm long), slender, spikelets closely imbricate (usually ½), solitary at each node

spikelets   4- to 7-flowered (1–2 cm long); rachilla pubescent

awns   lemmas awned, short or long (may be awnless); glumes may taper to short awns

glumes   broad, nearly enclose the florets, strongly nerved, nerves dark green, scabrous and hyaline margins

VEGETATIVE CHARACTERISTICS

culm   erect (0.5–1.0 m tall), tufted, dark nodes, green or glaucous, glabrous

sheath   round, glabrous, open; auricles sometimes present, 1 auricle often rudimentary

blade   slender with pointed tips (5–25 cm long and 2–7 mm wide), glabrous; margin with narrow white band, slightly barbed

ligule   membranous (0.4–0.8 mm long), rounded, lobed

GROWTH CHARACTERISTICS   starts growth in mid-spring, seeds mature by August–September, reproduces by seeds and tillers

FORAGE VALUE   excellent for both sheep and cattle when green and good when mature, good to excellent for wildlife

HABITAT   moist to well drained soils along rivers and in meadows, as well as under open forest canopy; can grow in moderate alkaline soils

Canada wildrye (*Elymus canadensis*)

Notes:

198

Tribe	TRITICEAE
Species	*Elymus canadensis* L.
Common Name	Canada wildrye
Life Span	Perennial
Origin	Native
Season	Cool

INFLORESCENCE CHARACTERISTICS

type   spike (8–25 cm long), erect or nodding, somewhat elliptical, thick, bristly; 2–4 spikelets per node, slightly spreading

spikelets   2- to 6-flowered; lemmas broad at base (0.8–1.0 mm long), hispid or scabrous

awns   lemma awned (1.5–5.0 cm long), curving outward at maturity

glumes   narrow, about equal, tapering to an awn (2–3 cm long)

VEGETATIVE CHARACTERISTICS

culm   decumbent at base (1.0–1.5 m tall), tufted, coarse

sheath   glabrous, rarely pubescent; auricles well developed (1–2 mm long), clasping, pointed

blade   elongate (5–40 cm long and 0.7–2.0 cm wide), flat or folded, tapering to a fine point, scabrous above, midrib prominent beneath, margin finely toothed

ligule   membranous (0.5–1.0 mm long), truncate, ciliate

GROWTH CHARACTERISTICS   starts growth in fall and makes some growth in winter, seeds mature by late spring–early summer, reproduces by seeds and tillers

FORAGE VALUE   good for cattle and horses, fair for sheep and wildlife during the spring when green and growing, value decreases sharply when the plant matures

HABITAT   prairies, moist sites, and disturbed areas

Basin wildrye (*Elymus cinereus*)

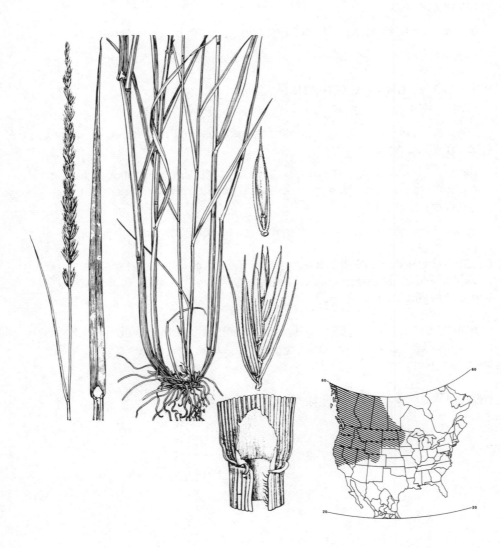

Tribe	TRITICEAE
Species	*Elymus cinereus* Scribn. and Merr.
Common Name	Basin wildrye (giant wildrye)
Life Span	Perennial
Origin	Native
Season	Cool

INFLORESCENCE CHARACTERISTICS

type   spike (10–25 cm long), thick, dense, generally erect, typically not branched; 3–5 spikelets per node (1 may be pedicellate)

spikelets   3- to 6-flowered (1.5 cm long); lemmas glabrous to sparsely strigose, with hyaline margin, awnless or mucronate

glumes   narrow, subulate, awn-pointed, 1-nerved or nerveless, about as long as first lemma, sometimes longer

VEGETATIVE CHARACTERISTICS

culm   erect (1–3 m tall), coarse, robust, harsh-puberulent at least above the nodes

sheath   glabrous to densely harsh-puberulent, closed below the collar; auricles finger-like (1–2 mm long)

blade   flat to involute (20–60 cm long and 0.5–1.5 cm wide), narrowing to an acute tip, glabrous to harsh-puberulent

ligule   membranous (5–7 mm long), rounded to acute, erose

other   rarely with short rhizomes

GROWTH CHARACTERISTICS   starts growth in early spring, seeds mature by August, reproduces by seed and tillers

FORAGE VALUE   good for cattle and fair for sheep and wildlife, relatively unpalatable in the summer

HABITAT   river banks, ravines, moist or dry slopes, and plains

Foxtail barley (*Hordeum jubatum*)

Tribe	TRITICEAE
Species	*Hordeum jubatum* L.
Common Name	Foxtail barley
Life Span	Perennial
Origin	Native
Season	Cool

INFLORESCENCE CHARACTERISTICS

type   spicate raceme (4–10 cm long and 4–6 cm wide, including awns), nodding, often purple; 3 spikelets per node, 1 fertile and sessile, 2 lateral spikelets pedicellate and sterile

spikelets   1-flowered, florets of lateral reduced spikelets usually equalling or exceeding the central spikelet

awns   lemma and glumes awned (1–6 cm long), glume awns scabrous

glumes   outer glumes of lateral spikelet narrow, setaceous; glumes of central spikelet and inner glumes of lateral spikelets broadened and flattened below, coarse-ciliate on margins

other   inflorescence may be partially enclosed in upper leaf sheath

VEGETATIVE CHARACTERISTICS

culm   erect or decumbent (30–75 cm tall), nodes dark

sheath   round, glabrous

blade   flat (5–15 cm long and 2–5 mm wide), tapering to fine point, smooth on upper surface

ligule   membranous (0.2–1.0 mm long), truncate

GROWTH CHARACTERISTICS   starts growth in late April–May, matures June–August, reproduces by seeds and tillers

FORAGE VALUE   poor for all classes of livestock and wildlife, it may be lightly grazed before inflorescence development

HABITAT   open ground, meadows, and waste places

Little barley (*Hordeum pusillum*)

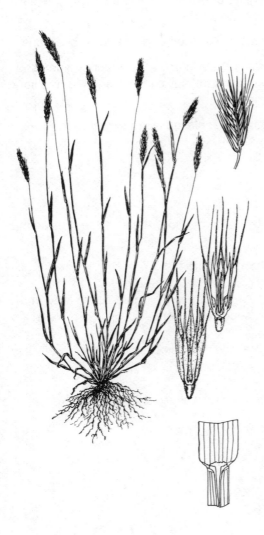

Tribe	TRITICEAE
Species	*Hordeum pusillum* Nutt.
Common Name	Little barley
Life Span	Annual
Origin	Native
Season	Cool

INFLORESCENCE CHARACTERISTICS

type   spicate raceme (2–8 cm long excluding awns), erect, narrow, dense; 3 spikelets per node, 1 fertile and sessile, 2 lateral spikelets pedicellate and sterile

spikelets   1-flowered, lemmas of lateral spikelets ½ to ⅓ as long as the lemma of the central spikelet

awns   lemma of central spikelet awned (2–7 mm long), lemmas of lateral spikelets short-awned, outer glumes of lateral spikelets awn-like, other glumes awned (0.7–1.5 cm long)

glumes   dilated above the base, scabrous

VEGETATIVE CHARACTERISTICS

culm   tufted (10–40 cm tall), usually geniculate at base; nodes glabrous and dark

sheath   round, glabrous or with short spreading pubescence, generally with auricles

blade   flat (1–12 cm long and 2–5 mm wide), erect, margin weakly barbed, glabrous or pubescent

ligule   membranous (0.4–0.7 mm long), truncate, erose to short ciliate

GROWTH CHARACTERISTICS   starts growth in early spring, matures by May–June, reproduces by seed

FORAGE VALUE   essentially no value for either livestock or wildlife, although it is sometimes lightly grazed in the spring

HABITAT   plains, open ground, especially alkaline areas

Squirreltail (*Sitanion hystrix*)

Tribe	TRITICEAE
Species	*Sitanion hystrix* (Nutt.) J. G. Smith
Common Name	Squirreltail (bottlebrush squirreltail, zacate trigillo)
Life Span	Perennial
Origin	Native
Season	Cool

INFLORESCENCE CHARACTERISTICS

type   spike (2–10 cm long), partially enclosed in sheath, erect, 2 spikelets per node

spikelets   2- to 6-flowered; lemmas convex (0.8–1.0 cm long), glabrous to lightly pubescent

awns   lemmas awned (0.5–1.5 cm long); glumes awned (2–10 cm long), widely spreading, stiff

glumes   entire or bifid, narrow

VEGETATIVE CHARACTERISTICS

culm   erect to spreading (10–50 cm tall), densely tufted, stiff

sheath   translucent margin, open, pubescent or glabrous, glabrous collar; auricles small (1 mm long)

blade   flat to involute (5–20 cm long and 1–5 mm wide), tapering to a fine point, prominently veined, harsh, stiffly ascending

ligule   membranous (0.6–1.0 mm long), erose to ciliate, rounded

GROWTH CHARACTERISTICS   starts growth in early spring, may flower 2 times per year with favorable moisture, reproduces by seed

LIVESTOCK LOSSES   sharp seeds and awns may cause injury

FORAGE VALUE   fair for cattle and sheep before maturity of inflorescences, unpalatable during fall and winter

HABITAT   dry hills, plains, open woods, and rocky slopes

Medusahead rye (*Taeniatherum asperum*)

208

Tribe	TRITICEAE
Species	*Taeniatherum asperum* (Simonkai) Nevski
Common Name	Medusahead rye
Life Span	Annual
Origin	Introduced (from Europe)
Season	Cool

INFLORESCENCE CHARACTERISTICS

type   spike (2–5 cm long), nearly as wide as long, solitary, bristly, 2 spikelets per node

spikelets   1-flowered; lemma lanceolate (6 mm long), scabrous

awns   lemma awned (5–10 cm long), flat; glumes awned (1.0–2.5 cm long)

glumes   subulate, smooth, indurate below, tapering into an awn

VEGETATIVE CHARACTERISTICS

culm   decumbent and branching at the base (20–60 cm tall), weak

sheath   open, strigose-puberulent to glabrous

blade   involute (3–10 cm long and 1–3 mm wide), glabrous to puberulent

ligule   membranous (0.5 mm long), rounded

GROWTH CHARACTERISTICS   starts growth in March–April, matures May–June, a problem weed, reproduces by seeds

LIVESTOCK LOSSES   stiff awns have caused injury by working into the ears, eyes, nose, and tongue

FORAGE VALUE   poor for livestock and wildlife, may be grazed early in the spring

HABITAT   open ground, soils, disturbed areas, waste places

Skunkbrush sumac (*Rhus trilobata*)

Notes:

Family	ANACARDIACEAE
Species	*Rhus trilobata* Nutt.
Common Name	Skunkbrush sumac
Life Span	Perennial
Origin	Native
Season	Cool

GROWTH CHARACTERISTICS   shrub (0.5–2.5 m tall), deciduous, flowers before the leaves appear in early spring

DISTINGUISHING CHARACTERISTICS

leaves   compound, 3 leaflets each shallowly lobed (1–3 cm long), green above, pale below, alternate, felty pubescent, leaflets sessile or nearly so, spatula-shaped and wedge-shaped at the base, terminal leaflet longer

flowers   yellow to cream colored, inconspicuous, in clusters near branch tips early in season

fruit   red or reddish-orange drupe (6–7 mm long), glandular pubescent, in short dense clusters, persist in winter

other   plants may appear with no leaves at all, branches alternate, ill-scented

HISTORIC, FOOD, AND MEDICINAL USES   some tribes of American Indians made use of the berry-like fruits as food, medicine, and lemonade-like drinks, slender shoots were used for basket weaving

LIVESTOCK LOSSES   none

FORAGE VALUE   palatability is very poor to fair for cattle and sheep, good browse for wildlife and goats

HABITAT   dry rocky hillsides, to some extent along streams and canyon bottoms

Western waterhemlock (*Cicuta douglasii*)

Family	APIACEAE
Species	*Cicuta douglasii* (DC.) Coult. & Rose
Common Name	Western waterhemlock
Life Span	Perennial
Origin	Native
Season	Cool

GROWTH CHARACTERISTICS   forb (0.6–1.2 m tall), erect, growing in or near water from tuberous roots

DISTINGUISHING CHARACTERISTICS

leaves   once or twice pinnately compound, alternate, leaflets lanceolate (3–10 cm long), coarsely serrate, veins in the leaflets run to the sinuses between the teeth

flowers   5-petaled white or greenish, small, and numerous in compound umbels

fruit   ovoid to subglobose (2–4 mm long), ribbed

other   smooth stems; roots and lower stem have cross partitions inside forming horizontal chambers; tuberous roots

HISTORIC, FOOD, AND MEDICINAL USES   roots were eaten by Cherokee women for 4 consecutive days to induce permanent sterility

LIVESTOCK LOSSES   poisonous, contains cicutoxin (yellowish liquid oil), stems and leaves are poisonous when young, dried roots are extremely toxic and easily pulled up by cattle since it generally grows in moist soils

FORAGE VALUE   none

HABITAT   moist soil and shady stream banks

Yaupon holly (*Ilex vomitoria*)

Notes:

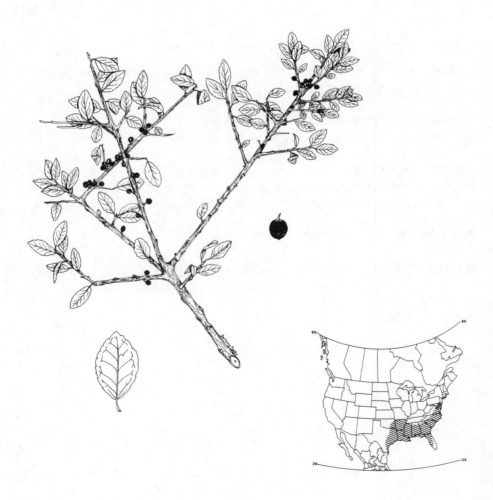

Family	AQUIFOLIACEAE
Species	*Ilex vomitoria* Ait.
Common Name	Yaupon holly (yaupon)
Life Span	Perennial
Origin	Native
Season	Evergreen

GROWTH CHARACTERISTICS   shrub or small tree (to 8 m tall), densely branched from a rounded crown, often forming dense thickets, flowers in April or May, reproduces from sprouts as well as from seeds

DISTINGUISHING CHARACTERISTICS

leaves   simple (2−4 cm long), alternate, petiolate (1−3 mm), elliptic-oblong to oval, broadest at middle, crenate margin, dark glands on margin, dark green lustrous above, paler beneath, puberulent above along midrib and at base, otherwise glabrous

flowers   white; unisexual, plants dioecious; staminate flowers in dense axillary clusters; pistillate flowers solitary or few

fruit   bright red drupe (5−8 mm long), 4-seeded, nutlets grooved

other   stem light gray to white; bark thin, light reddish-brown, twigs short, stout and rigid, obtusely angled, densely puberulent

HISTORIC, FOOD, AND MEDICINAL USES   Indians made bitter tea called "black drink" from the leaves, which contain caffeine, for ceremonial reasons; used as a holiday decoration

LIVESTOCK LOSSES   none

FORAGE VALUE   fair browse for livestock; red fruits provide food for wild turkey, quail, and song birds; whitetail deer browse young twigs and leaves

HABITAT   streams and pond margins; shallow swamps, wet woods, and sandy pinelands

Western yarrow (*Achillea millefolium*)

Family	ASTERACEAE
Tribe	ANTHEMIDEAE
Species	*Achillea millefolium* L.
Common Name	Western yarrow
Life Span	Perennial
Origin	Native
Season	Cool

GROWTH CHARACTERISTICS   forb (0.2–1.0 m tall), erect, branching above, extensive horizontal rootstalks, long flowering periods

DISTINGUISHING CHARACTERISTICS

leaves   simple, alternate, finely dissected (fern-like), ultimate divisions linear, villous

flowers   white, radiate; several heads, forming a flat-topped inflorescence

fruit   achenes with no pappus, flat and chaff-like

other   pubescent stem; strong odor

HISTORIC, FOOD, AND MEDICINAL USES   according to tradition, the plant was first named by Achilles, hence its name; decoction of leaves and flowers was used by Blackfoot Indians as an eyewash; Winnebagos steeped whole plants and poured the liquid into aching ears; green leaves were used to relieve itching, chewed for toothaches, and used as mild laxatives; leaves boiled for tea were used as a cold remedy

LIVESTOCK LOSSES   not generally considered to be poisonous, but may contain alkaloid and glycoside toxic principles

FORAGE VALUE   poor to rarely good for cattle, sheep may graze it during the growing season; frequently scattered where ranges have been heavily grazed

HABITAT   sagebrush areas, canyon bottoms, glades, pastures, and roadsides; prevalent in open timber and disturbed sites

Silver sagebrush (*Artemisia cana*)

Family	ASTERACEAE
Tribe	ANTHEMIDEAE
Species	*Artemisia cana* Pursh
Common Name	Silver sagebrush (white sagebrush)
Life Span	Perennial
Origin	Native
Season	Warm

GROWTH CHARACTERISTICS   shrub (1.0–1.5 m tall), rounded, sometimes form extensive colonies by rhizomes, stems much-branched; flowers September, fruits October–November, reproduces by seeds and rhizomes

DISTINGUISHING CHARACTERISTICS

leaves   simple, alternate, linear, entire or occasionally with 1–2 irregular teeth, canescent, may appear fascicled on short lateral branches

flowers   discoid, in a leafy panicle; heads in groups; 8–15 papery phyllary bracts

fruit   achene, 4–5 ribbed

other   plants often dry with a goldfish hue and yellowish stem

HISTORIC, FOOD, AND MEDICINAL USES   decoction used by American Indians to stop coughing, extract of plant was taken to restore hair

LIVESTOCK LOSSES   none

FORAGE VALUE   important browse for livestock; good to excellent in fall and winter; increases under cattle and decreases under sheep browsing; some varieties are not palatable

HABITAT   river valley terraces and uplands, dry soils, deep loam to sandy soils

Sand sagebrush (*Artemisia filifolia*)

Family	ASTERACEAE
Tribe	ANTHEMIDEAE
Species	*Artemisia filifolia* Torr.
Common Name	Sand sagebrush
Life Span	Perennial
Origin	Native
Season	Warm

GROWTH CHARACTERISTICS   shrub (generally less than 1 m tall), rounded, freely branched, growth starts in early May, flowers until the fall frost, reproduces by seeds

DISTINGUISHING CHARACTERISTICS

leaves   alternate (3.5–7.5 cm long), sessile, often fascicled, thread-like, upper leaves entire, lower in narrow divisions, 3-parted

flowers   in dense leafy panicles, small discoid heads

fruit   achene without pappus

other   fine-stemmed; twigs often striate; loses pubescence with age

HISTORIC, FOOD, AND MEDICINAL USES   in Mexico, a decoction of leaves was taken for intestinal worms and other stomach problems

LIVESTOCK LOSSES   may cause sage sickness in horses

FORAGE VALUE   poor to worthless to cattle, poor to fair for horses and sheep; furnishes forage for antelope and deer

HABITAT   sandy soil, generally considered an indicator of sandy soil

Fringed sagebrush (*Artemisia frigida*)

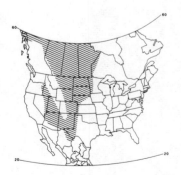

Family	ASTERACEAE
Tribe	ANTHEMIDEAE
Species	*Artemisia frigida* Willd.
Common Name	Fringed sagebrush (prairie sagewort, wormwood)
Life Span	Perennial
Origin	Native
Season	Cool

GROWTH CHARACTERISTICS   suffrutescent shrub (10–40 cm tall), often forming mats with numerous upright stems, increases rapidly under heavy grazing, drought tolerant

DISTINGUISHING CHARACTERISTICS

leaves   appear fascicled (0.5–1.5 cm long), dissected 3–5 times into linear divisions, silvery-canescent, crowded below, upper leaves may be dissected

flowers   discoid (2.0–3.5 mm long and 4–6 mm wide), yellowish, in leafy racemes; bracts long-hairy

fruit   achene

other   entire plant is silky-canescent; aromatic plants may turn brown on drying

HISTORIC, FOOD, AND MEDICINAL USES   American Indians called it "woman sage" and used it to eliminate the greasy smell from dried meat; to bandage cuts after it was chewed; to make mats, fans, menstrual pads; as toilet paper

LIVESTOCK LOSSES   none

FORAGE VALUE   good for sheep and goats, poor to fair for cattle; important winter feed for elk and deer

HABITAT   dry prairies and plains, hills and mountains; favors porous soil

Cudweed sagewort (*Artemisia ludoviciana*)

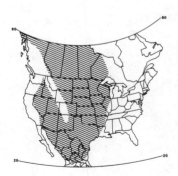

Family	ASTERACEAE
Tribe	ANTHEMIDEAE
Species	*Artemisia ludoviciana* Nutt.
Common Name	Cudweed sagewort (Louisiana wormwood, western mugwort)
Life Span	Perennial
Origin	Native
Season	Warm

GROWTH CHARACTERISTICS   forb (30–90 cm tall), erect, often growing in colonies from creeping rootstalks

DISTINGUISHING CHARACTERISTICS

leaves   alternate, variable from linear to oblanceolate (2.5–10.0 cm long and usually less than 1 cm wide), reduced upwards, margins entire to toothed, dense, white-tomentose, leaves are often recurved when plants mature

flowers   in small discoid heads (3–4 mm wide), gray or white-woolly

fruit   achenes (less than 1 mm long), no pappus

other   entire plant is aromatic

HISTORIC, FOOD, AND MEDICINAL USES   American Indians called it "man sage" and used it for ceremonial and purification purposes; to deodorize feet; cure headaches, treat coughs, hemorrhoids, stomach troubles, and wounds on horses; made into pillows and saddle pads; burned to drive mosquitos away

LIVESTOCK LOSSES   none

FORAGE VALUE   generally fair to poor, decreasing from South to North; somewhat palatable to elk and deer

HABITAT   prairies and roadsides, foothills, and plains

Black sagebrush (*Artemisia nova*)

226

Family	ASTERACEAE
Tribe	ANTHEMIDEAE
Species	*Artemisia nova* A. Nels.
Common Name	Black sagebrush
Life Span	Perennial
Origin	Native
Season	Evergreen

GROWTH CHARACTERISTICS   shrub (10–40 cm tall), low, spreading, flowers September–November, reproduces by seeds

DISTINGUISHING CHARACTERISTICS

leaves   alternate, wedge-shaped to fan-shaped (0.5–1.5 cm long and 0.3–1.0 cm wide), 3- to 5-lobed at apex, resinous before flowering, canescent

flowers   inflorescence narrow, golden brown; heads small, discoid; phyllary bracts 8–12, dry, imbricate, inner series glabrous

fruit   achenes, glabrous

other   dark brown to black bark; twigs canescent becoming glabrate

HISTORIC, FOOD, AND MEDICINAL USES   some American Indians drank a decoction of the boiled stems, leaves, and twigs for bronchitis; leaves were crushed and the vapor inhaled to relieve nasal congestion

LIVESTOCK LOSSES   none

FORAGE VALUE   poor for cattle and fair for sheep, cattle will use it in fall and winter, excellent wildlife browse

HABITAT   dry slopes and ridges, rocky places

Budsage (*Artemisia spinescens*)

Family	ASTERACEAE
Tribe	ANTHEMIDEAE
Species	*Artemisia spinescens* (DC.) Eat.
Common Name	Budsage (bud sagewort)
Life Span	Perennial
Origin	Native
Season	Cool

GROWTH CHARACTERISTICS   shrub (10–60 cm tall), rounded, much-branched, rigid, spinescent, leaves form in early spring, flowers March–June, reproduces by seeds

DISTINGUISHING CHARACTERISTICS

leaves   simple (0.5–1.5 cm long), alternate, fan-shaped in outline, 3–5 parted, woolly

flowers   on short leafy bracted branches which become spinescent twigs

fruit   achene, densely long-hairy

other   spinescent, very little herbaceous growth

HISTORIC, FOOD, AND MEDICINAL USES   pollen causes hay fever, indicator of alkaline soils

LIVESTOCK LOSSES   thought to be poisonous when not browsed with other plants

FORAGE VALUE   valuable early browse for sheep in Utah, Nevada, and California; resistant to browsing

HABITAT   desert mesas and plains, dry sunny sites; alkaline soils, usually in association with shadscale (*Atriplex confertifolia*)

Big sagebrush (*Artemisia tridentata*)

Notes:

Family	ASTERACEAE
Tribe	ANTHEMIDEAE
Species	*Artemisia tridentata* Nutt.
Common Name	Big sagebrush
Life Span	Perennial
Origin	Native
Season	Evergreen

GROWTH CHARACTERISTICS   shrub or small tree (to 6 m tall), rounded crown, trunk short, evergreen to late-deciduous

DISTINGUISHING CHARACTERISTICS

leaves   alternate, simple, widge-shaped (1–4 cm long and 0.3–1.4 cm wide), usually 3-lobed at apex, dense gray hairs on both sides, pungent odor when crushed

flowers   in panicles, heads numerous

fruit   achenes, resinous

other   bark thin, shredding into long flat thin strips grayish-brown

HISTORIC, FOOD, AND MEDICINAL USES   firewood, pollen causes hay fever, decoctions used by some American Indians as a laxative, wood once used for thatch

LIVESTOCK LOSSES   none

FORAGE VALUE   grazed extensively by sheep on winter ranges, high in protein but also high in volatile oils; food source and cover for wildlife; leaves, fruits, and flowers are main diet of sage grouse, lesser degree for sharp-tailed and dusky grouse; browse for antelope, mule deer, elk, and mountain sheep; jackrabbits, ground squirrels, and rodents eat leaves and seeds

HABITAT   dry, gravelly, rocky soils; plains, deserts, hills, and lower mountain slopes

Rubber rabbitbrush (*Chrysothamnus nauseosus*)

Family	ASTERACEAE
Tribe	ASTEREAE
Species	*Chrysothamnus nauseosus* (Pall.) Britt.
Common Name	Rubber rabbitbrush (gray rabbitbrush)
Life Span	Perennial
Origin	Native
Season	Warm

GROWTH CHARACTERISTICS   shrub (0.5–1.0 m tall), deciduous, several erect branches from base forming a rounded clump, root sprouter

DISTINGUISHING CHARACTERISTICS

leaves   simple, alternate, sessile, linear (3–5 cm long and 0.5–4.0 mm wide), densely hairy to subglabrous, 1 prominent midvein, leaves seldom twisted

flowers   in rounded cymes or thyrses, small discoid heads (0.6–1.3 cm long); phyllary bracts graduated and tend to be ranked in 2–3 series

fruit   achene, linear and strigose; pappus of prominent capillary bristles, dull-white or tawny

other   stem tomentose, felt-like; twigs often ill-scented ("nauseous" odor); bark fibrous

HISTORIC, FOOD, AND MEDICINAL USES   American Indians made chewing gum from pulverized wood and bark, also used as tea, cough syrup, yellow dye, and for chest pains; small commercial source for rubber extraction

LIVESTOCK LOSSES   toxic to livestock

FORAGE VALUE   little or no value to livestock, fair for deer on winter range; dense stands may indicate poor range management

HABITAT   open dry places

Douglas rabbitbrush (*Chrysothamnus viscidiflorus*)

Notes:

Family	ASTERACEAE
Tribe	ASTEREAE
Species	*Chrysothamnus viscidiflorus* (Hook.) Nutt.
Common Name	Douglas rabbitbrush (green rabbitbrush)
Life Span	Perennial
Origin	Native
Season	Warm

GROWTH CHARACTERISTICS
shrub (0.5–1.0 m tall), round-topped, branched from base, reproduces by seeds and sprouts

DISTINGUISHING CHARACTERISTICS

leaves simple (1–6 cm long and 0.1–1.0 cm wide), alternate, linear-oblanceolate, glabrous or finely puberulent, flat or sometimes twisted, punctate glands may be present below, margins scabrous

flowers borne in terminal cymes; involucre, yellow (5–6 mm long); phyllary bracts graduated in poorly defined rows

fruit achenes sparsely pubescent (3–4 mm long), 5-angled; pappus stiff, white, at least as long as the corolla

other young branches are puberulent, stems and twigs glabrous

HISTORIC, FOOD, AND MEDICINAL USES
roots chewed as gum in the Southwest; contains rubber, especially when growing in alkali soils

LIVESTOCK LOSSES
none

FORAGE VALUE
poor, occasionally browsed by sheep and cattle when other feed is not available, deer browse lightly in the summer and winter, elk utilize it in winter

HABITAT
dry open places in sagebrush, ponderosa pine, lodgepole pine, or aspen belts

Curlycup gumweed (*Grindelia squarrosa*)

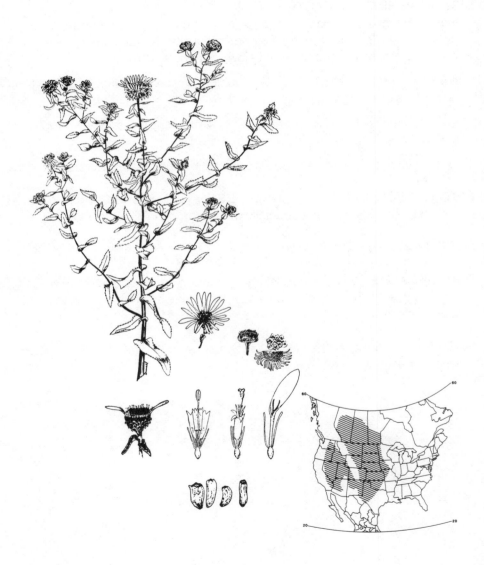

Family	ASTERACEAE
Tribe	ASTEREAE
Species	*Grindelia squarrosa* (Pursh) Dun.
Common Name	Curlycup gumweed (gumweed)
Life Span	Perennial
Origin	Native
Season	Warm

GROWTH CHARACTERISTICS   forb (0.2–1.0 m tall), erect, from a tap-root, much-branched above, starts growth in late spring, flowers July–August, reproduces by seeds

DISTINGUISHING CHARACTERISTICS

leaves   alternate, mostly oblong (2–5 cm long), serrulate with about 8 gland-dotted teeth on each side (rarely entire), thick and viscid from small glandular dots covering leaf surfaces, sessile to somewhat clasping

flowers   showy yellow heads (2–3 cm wide), generally radiate, however rays may be lacking; phyllary bracts obviously imbricated with their tips curled back (squarrose) and covered with sticky resin, hence the common name

fruit   glabrous achenes; pappus of stiff deciduous awns

HISTORIC, FOOD, AND MEDICINAL USES   American Indians used gummy secretions to relieve asthma, bronchitis and colic; Pawnee Indians boiled leaves and flowering tops to treat saddle sores and raw skin; flower extract is used in modern medicine to treat whooping cough and asthma

LIVESTOCK LOSSES   may accumulate selenium

FORAGE VALUE   seldom eaten by livestock due to stickiness

HABITAT   waste places, overgrazed prairies, and alluvial grounds

Broom snakeweed (*Gutierrezia sarothrae*)

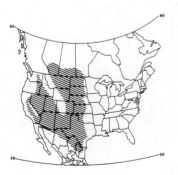

Family	ASTERACEAE
Tribe	ASTEREAE
Species	*Gutierrezia sarothrae* (Pursh) Britt. & Rusby
Common Name	Broom snakeweed (perennial broomweed, escobilla)
Life Span	Perennial
Origin	Native
Season	Warm

GROWTH CHARACTERISTICS   suffrutescent shrub (10–60 cm tall), bushy, much-branched

DISTINGUISHING CHARACTERISTICS

leaves   alternate, linear to filiform (2–4 cm long and 1–2 mm wide), margins entire, rolled inward, pubescent to scabrous, leafless below

flowers   inflorescence cymose, cylindric; heads yellow, radiate; phyllary bracts imbricate, may be green-tipped

fruit   achene, pappus of chaffy scales

other   stems slender; herbaceous above and woody below

HISTORIC, FOOD, AND MEDICINAL USES   used by southwestern Indians and Mexicans as a broom; decoctions used for indigestion

LIVESTOCK LOSSES   poisonous to sheep and cattle, causing death or abortion; poisonous principle is a saponin

FORAGE VALUE   generally considered worthless except on winter range where it is fair for sheep and poor for cattle and horses; indicator of overgrazing

HABITAT   rocky plains and dry hillsides; on nearly all soil types

Hairy goldaster (*Heterotheca villosa*)

Notes:

Family	ASTERACEAE
Tribe	ASTEREAE
Species	*Heterotheca villosa* (Pursh) Shinners
Common Name	Hairy goldaster (telegraph plant, golden aster)
Life Span	Perennial
Origin	Native
Season	Warm

GROWTH CHARACTERISTICS   forb (20–50 cm tall), from woody caudex or woody taproot, several erect or ascending stems, reproduces by seeds

DISTINGUISHING CHARACTERISTICS

leaves   alternate, reduced upward, lower leaves petiolate and spatulate to oblanceolate; middle leaves oblanceolate (1–3 cm long); upper leaves sessile, linear to oblanceolate, hirsute

flowers   corymbiform to cymose-paniculate, heads 1–30 per stem; large ray and disk flowers yellow-orange; phyllary bracts linear, imbricated in 4–9 series, hirsute and green-striped

fruit   achenes; pappus in 2 series, inner series of capillary bristles, outer of scales

other   entire plant appearing grayish-green due to pubescence which is generally pustulate-based hirsute to hispid; this species may have much variation due to several intergrating races and forms

HISTORIC, FOOD, AND MEDICINAL USES   some American Indians consumed a decoction from the tops and stems as a soothing, quieting medicine to aid sleep

LIVESTOCK LOSSES   none

FORAGE VALUE   fair for sheep in poor semidesert areas of the West, otherwise considered worthless

HABITAT   sandy and rocky soils, thriving in dry areas

Prairie goldenrod (*Solidago missouriensis*)

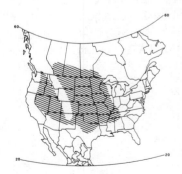

Family	ASTERACEAE
Tribe	ASTEREAE
Species	*Solidago missouriensis* Nutt.
Common Name	Prairie goldenrod (Missouri goldenrod)
Life Span	Perennial
Origin	Native
Season	Warm

GROWTH CHARACTERISTICS   forb (0.2–1.0 m tall), reproduces by seeds and rhizomes, flowers June–July, matures August–September

DISTINGUISHING CHARACTERISTICS

leaves   alternate, oblanceolate to linear, lower leaves largest, generally persistent, gradually reduced in size up the stem, taper to a petiole, margins entire to slightly toothed, 3 prominent midveins, glabrous but with rough margins

flowers   paniculiform, with recurved branches, generally 1-sided; heads small (less than 1 cm), numerous; ray and disk flowers yellow; phyllary bracts in a few graduated series

fruit   achene (1–2 mm long), rounded, pappus of capillary bristles

other   stem essentially glabrous

HISTORIC, FOOD, AND MEDICINAL USES   some American Indians chewed leaves and flowers to relieve sore throats, roots were chewed to relieve toothache; pollen highly desirable by several bee species

LIVESTOCK LOSSES   may be toxic to sheep

FORAGE VALUE   poor, however will be grazed by cattle and sheep in spring and early summer; generally becomes more abundant in overgrazed areas

HABITAT   dry open prairies and plains

243

Annual broomweed (*Xanthocephalum dracunculoides*)

Family	ASTERACEAE
Tribe	ASTEREAE
Species	*Xanthocephalum dracunculoides* (DC.) Shinners
Common Name	Annual broomweed (common broomweed, escobilla)
Life Span	Annual
Origin	Native
Season	Warm

GROWTH CHARACTERISTICS   forb (20−60 cm tall), simple below and much-branched above, therefore appearing bushy, flowers September–November, reproduces by seeds

DISTINGUISHING CHARACTERISTICS

leaves   alternate, linear to narrow elliptic (1.8−7.0 cm long and 0.5−5.0 mm wide), resinous, glabrous

flowers   heads often crowded; involucre (3−5 mm long); ray flowers yellow, fertile; disk flowers yellow, sterile; phyllary bracts firm, straw-colored with green tips

fruit   achene (1.5 mm long), pappus of slender membranous scales emerge as 5 linear strap-shaped white projections as long as the corolla

other   pungent odor

HISTORIC, FOOD, AND MEDICINAL USES   none

LIVESTOCK LOSSES   herbage and pollen can cause itchy dermatitis in livestock (and humans)

FORAGE VALUE   unpalatable to cattle; increases in abundance under heavy grazing pressure and during dry conditions

HABITAT   dry upland prairies, limestone barrens, roadsides, and railroad rights-of-way

False dandelion (*Agoseris glauca*)

Notes:

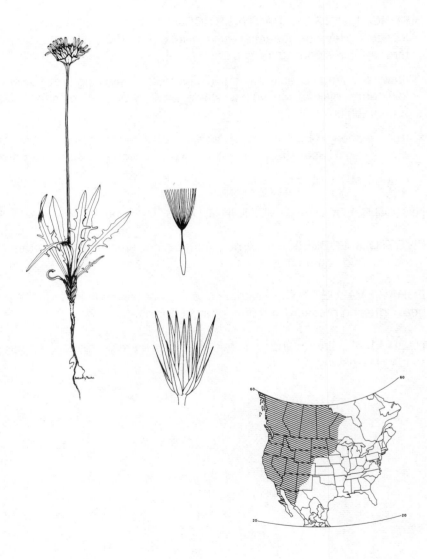

Family	ASTERACEAE
Tribe	CICHORIEAE
Species	*Agoseris glauca* (Pursh) Raf.
Common Name	False dandelion (mountain dandelion, pale agoseris)
Life Span	Perennial
Origin	Native
Season	Cool

GROWTH CHARACTERISTICS   forb (10–60 cm tall), acaulescent, tap-root

DISTINGUISHING CHARACTERISTICS

leaves   basal rosette, linear to oblanceolate (5–20 cm long), entire to pinnatifid, distinct white midrib, glabrous to pubescent

flowers   ligulate, yellow sometimes turning rose-purple with age, head solitary (2–4 cm wide), on a stout flowering scape (20–60 cm tall), petals are 5-toothed at their apex; phyllary bracts are erect, imbricate in 2 rows, often with purple spots or tips

fruit   achene (0.5–1.0 cm long), beaked, pappus capillary

other   milky sap

HISTORIC, FOOD, AND MEDICINAL USES   sap was chewed by American Indians to clean teeth

LIVESTOCK LOSSES   none

FORAGE VALUE   fair to good for sheep and deer, lightly grazed by cattle and horses; abundance indicates poor range condition

HABITAT   moist meadows, prairie swales; common on disturbed or eroded sites

Tapertip hawksbeard (*Crepis acuminata*)

Family	ASTERACEAE
Tribe	CICHORIEAE
Species	*Crepis acuminata* Nutt.
Common Name	Tapertip hawksbeard (hawksbeard)
Life Span	Perennial
Origin	Native
Season	Cool

GROWTH CHARACTERISTICS   forb (20–65 cm tall), stems simple or 1–2 forked, slender stout woody taproot, caudex usually covered with brown bases of old leaves, flower May–August

DISTINGUISHING CHARACTERISTICS

leaves   mostly basal (12–40 cm long and 0.5–11.0 cm wide), pinnately lobed with 5–10 pairs of lateral segments, outline broadly lanceolate, apical segment (3–8 cm long) gradually attenuate acuminate, lobes entire or denate, blade narrows to narrowly-winged stout petiole with scarious base, densely or sparsely gray-green tomentose; cauline leaves remote

flowers   inflorescence a flat-topped cyme; heads numerous; flowers ligulate, yellow, small (0.9–1.5 cm long and 2.5–4.0 mm wide), 5–12 flowered; 5–7 outer phyllary bracts lance-deltoid and ciliate on margins, 5–8 inner phyllary bracts ciliate at apex

fruit   achene, pale yellow to brownish (5–9 mm long); pappus dusky white, bristle-like, united at base forming clumps

other   stems striate or sulcate, tomentose near base, branched from near or above middle; plants contain milky juice

HISTORIC, FOOD, AND MEDICINAL USES   none

LIVESTOCK LOSSES   none

FORAGE VALUE   grazed by all classes of livestock, preferred forage of sheep

HABITAT   dry, well drained areas, or shallow soils

Dandelion (*Taraxacum officinale*)

Family	ASTERACEAE
Tribe	CICHORIEAE
Species	*Taraxacum officinale* Weber
Common Name	Dandelion (common dandelion)
Life Span	Perennial
Origin	Introduced (from Eurasia)
Season	Cool

GROWTH CHARACTERISTICS   forb (5–30 cm tall), fleshy taproot, starts growth in early spring, reproduces by seeds

DISTINGUISHING CHARACTERISTICS

leaves   crowded in a basal rosette (5–25 cm long), variously deeply cut to subentire, pointed lobe tips generally point towards base of plant

flowers   ligulate; involucre yellow (1–2 cm wide), solitary on flowering scape (3–20 cm tall); phyllary bracts in 2 series, linear, outer series shorter and turned downward early, inner series turned upward at maturity

fruit   achene, slightly flattened, anterior portion often with antrorse-spinose projections, long filiform beak tipped with pappus of several white capillary bristles

other   deep taproot, reproduces by sexual seeds and apomixis

HISTORIC, FOOD, AND MEDICINAL USES   young leaves can be eaten as spring greens, roots can be ground and used as mild laxative or to treat heartburn; good honey plant; tea and wine can be made from the flowers

LIVESTOCK LOSSES   none

FORAGE VALUE   readily eaten by all livestock since it is relatively succulent; generally abundant on overgrazed range, but can also occur on well managed rangeland

HABITAT   common on rangeland, weedy meadows, and open stream banks; on a wide variety of soils

Dotted gayfeather (*Liatris punctata*)

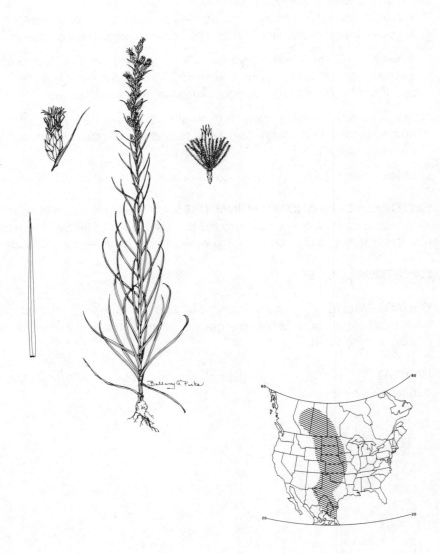

Family	ASTERACEAE
Tribe	EUPATORIEAE
Species	*Liatris punctata* Hook.
Common Name	Dotted gayfeather (blazing star)
Life Span	Perennial
Origin	Native
Season	Warm

GROWTH CHARACTERISTICS   forb (15–80 cm tall), 1-several stems, flowers August–October, reproduces by seeds and from a bulb-like caudex

DISTINGUISHING CHARACTERISTICS

leaves   alternate, linear, rigid, numerous, basal leaves (8–15 cm long and 1.5–6.0 mm broad), gradually becoming smaller up the stem, margin white ciliate-scabrous, surfaces punctate and glabrous

flowers   heads in dense spike-like arrangements (6–30 cm long), heads (1.5–2.0 cm long) discoid with 4–8 purple flowers; phyllary bracts thick, punctate, and bearing prominent ciliate-margins

fruit   ribbed, pubescent achenes (6–7 mm long); pappus plumose, often exceeding corolla

other   elongated bulb-like base (corm)

HISTORIC, FOOD, AND MEDICINAL USES   bulb reportedly used by American Indians for food; plants of this genus were consumed in New England for treatment of gonorrhea

LIVESTOCK LOSSES   none

FORAGE VALUE   grazed by domestic livestock, particularly sheep, especially when plants are young; disappears with continuous overuse

HABITAT   dry plains and hills

Triangleleaf bursage (*Ambrosia deltoidea*)

Family	ASTERACEAE
Tribe	HELIANTHEAE
Species	*Ambrosia deltoidea* (Torr.) Payne
Common Name	Triangleleaf bursage (triangle bursage)
Life Span	Perennial
Origin	Native
Season	Cool

GROWTH CHARACTERISTICS   shrub (to 1 m tall) with erect branches from a woody base, grows in almost pure stands with most growth from January—April, reproduces by seeds

DISTINGUISHING CHARACTERISTICS

leaves   alternate, distinctly petioled, ovate to mostly deltoid-shaped, green above and densely canescent below, appearing whitish-gray and velvety, margins serrate to serrulate

flowers   unisexual, plants monoecious; staminate heads in terminal racemes; pistillate below in a spiny involucre; ray flowers yellow with black glands

fruit   achene, rounded or ovate (6 mm log), 2—3 beaked at apex, glandular and tomentose with 20 or more flattened often hooked spines

HISTORIC, FOOD, AND MEDICINAL USES   none

LIVESTOCK LOSSES   fruit may get into fleece and reduce its value

FORAGE VALUE   worthless

HABITAT   hillsides, mesas, plains, and gullies

White bursage (*Ambrosia dumosa*)

Notes:

Family	ASTERACEAE
Tribe	HELIANTHEAE
Species	*Ambrosia dumosa* (Gray) Payne
Common Name	White bursage (burroweed)
Life Span	Perennial
Origin	Native
Season	Warm

GROWTH CHARACTERISTICS   shrub (to 1 m tall), much-branched, compact, spinescent, growth period April–November, reproduces by seeds

DISTINGUISHING CHARACTERISTICS

leaves   alternate or fascicled, 1–3 times pinnately divided into ovate or obovate divisions, canescent-pubescent (especially below) giving a gray to white color

flowers   unisexual, plants monoecious; staminate heads in terminal racemes with pistillate heads below, greenish-yellow, small

fruit   achene (4–6 mm long), bearing 25–40 rigid, flattened, straight spines

other   stems white

HISTORIC, FOOD, AND MEDICINAL USES   none

LIVESTOCK LOSSES   may accumulate nitrates

FORAGE VALUE   fair browse plant for cattle and horses, fair to good for goats

HABITAT   dry plains and mesas

Western ragweed (*Ambrosia psilostachya*)

Family	ASTERACEAE
Tribe	HELIANTHEAE
Species	*Ambrosia psilostachya* DC.
Common Name	Western ragweed
Life Span	Perennial
Origin	Native
Season	Warm

GROWTH CHARACTERISTICS   forb (30–80 cm tall), erect, forms extensive colonies from creeping horizontal rootstalks, also reproduces by seeds

DISTINGUISHING CHARACTERISTICS

leaves   opposite (4–12 cm long), deeply narrow-lobed to linear-lanceolate divisions; covered with short, stiff hairs with pustulate bases; sessile or nearly so

flowers   unisexual, plants monoecious; small and greenish, male flowers in terminal racemes; anthers yellow; female flowers axillary; inflorescence much-branched

fruit   seed with a woody hull with pointed tip surrounded by short spines

HISTORIC, FOOD, AND MEDICINAL USES   leaves were steeped by American Indians and used as a treatment for sore eyes; pollen causes hay fever

LIVESTOCK LOSSES   may accumulate nitrates, milk produced from cows grazing this forb has a bitter taste

FORAGE VALUE   generally unpalatable, cattle may graze in late summer and early spring

HABITAT   dry prairies, barrens, loess hills, pastures, roadsides and along railroads

Desert marigold (*Baileya multiradiata*)

Family	ASTERACEAE
Tribe	HELIANTHEAE
Species	*Baileya multiradiata* Harv. & Gray *ex* Torr.
Common Name	Desert marigold (paperdaisy, desert baileya)
Life Span	Perennial (or biennial)
Origin	Native
Season	Warm

GROWTH CHARACTERISTICS   forb (10–50 cm tall), short-lived, basal leaves persist as a rosette and in the lower portion of the stem, flowering peduncle bractless, flowers April–July when sufficient moisture is available, reproduces by seeds

DISTINGUISHING CHARACTERISTICS

leaves   mostly basal (4–10 cm long), pinnately parted to divided (irregularly lobed), taper to a petiole, white or gray, floccose-woolly throughout

flowers   large, radiate heads; rays broadly linear, bright yellow, 3-toothed at the apex, becoming papery in texture and reflexed; disk flowers numerous, yellow, 5-toothed; pappus absent

fruit   flattened achenes, truncate

other   white pubescence throughout

HISTORIC, FOOD, AND MEDICINAL USES   cultivated as an ornamental

LIVESTOCK LOSSES   sheep and goats may be killed after consuming large quantities over a relatively long period of time, poisonous principle has not been determined

FORAGE VALUE   poor to worthless for domestic livestock and wildlife, generally will not be grazed if other forage plants are present

HABITAT   dry plains, sandy, and gravelly soils

Arrowleaf balsamroot (*Balsamorhiza sagittata*)

Family	ASTERACEAE
Tribe	HELIANTHEAE
Species	*Balsamorhiza sagittata* (Pursh) Nutt.
Common Name	Arrowleaf balsamroot (gray dock, breadroot)
Life Span	Perennial
Origin	Native
Season	Cool

GROWTH CHARACTERISTICS   forb (20–60 cm tall), scapose, starts growth in April, matures by mid-summer, reproduces by seeds

DISTINGUISHING CHARACTERISTICS

leaves   mostly in basal clumps on long petioles, blade sagittate or arrow-shaped (15–30 cm long and 5–15 cm wide), margins entire, surfaces tomentose (more so below)

flowers   heads radiate, large (6–10 cm across), usually solitary, yellow disc and ray flowers; phyllary bracts in 2–4 series

fruit   glabrous achenes

other   from a large resinous taproot

HISTORIC, FOOD, AND MEDICINAL USES   Cheyenne Indians boiled roots, stems, and leaves and drank the decoction for stomach pains and headaches; they also steamed the plant and inhaled the vapors for the same purposes; ripe seeds were pounded into flour; roots were commonly eaten raw or boiled

LIVESTOCK LOSSES   none

FORAGE VALUE   good for sheep and fair for cattle, good for big game; green material and flowers being the most palatable portion; deer and elk feed on leaves

HABITAT   flats and open hillsides

Tarbush (*Flourensia cernua*)

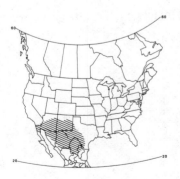

Family	ASTERACEAE
Tribe	HELIANTHEAE
Species	*Flourensia cernua* DC.
Common Name	Tarbush (hojase, blackbrush, varnishbush)
Life Span	Perennial
Origin	Native
Season	Warm

GROWTH CHARACTERISTICS　　shrub (1–2 m tall), leafy, highly branched, new growth starts in mid-summer, flowers in late fall

DISTINGUISHING CHARACTERISTICS

leaves　simple (1.7–2.5 cm long and 0.6–1.2 cm wide), alternate, elliptic, entire margin, petioled, sticky, covered with small black glands often causing plant to appear black when dried

flowers　yellow discoid; head nodding (1 cm wide); phyllary bracts in 3 series, linear, tips often spreading, black

fruit　laterally compressed achenes, pappus of 2 unequal ciliate awns

other　glabrous and glutinous, with an odor of tar; stem white scaly from coarse hairs and dried resin

HISTORIC, FOOD, AND MEDICINAL USES　　decoction is made in Mexico from leaves and flowers for indigestion

LIVESTOCK LOSSES　　may cause losses of sheep and goats January–March, following fruit maturity but before it falls from the plant

FORAGE VALUE　　livestock generally do not utilize this plant, increases with overgrazing; may be utilized by jackrabbits and other wildlife

HABITAT　　deserts and dry soils of valleys, mesas, flats, and foothills

Orange sneezeweed (*Helenium hoopesii*)

Family	ASTERACEAE
Tribe	HELIANTHEAE
Species	*Helenium hoopesii* Gray
Common Name	Orange sneezeweed (owlsclaws, western sneezeweed)
Life Span	Perennial
Origin	Native
Season	Warm

GROWTH CHARACTERISTICS   forb, from rhizome or woody caudex (0.4–1.0 m tall), few to several stems, flowers in July–August, reproduces by seeds

DISTINGUISHING CHARACTERISTICS

leaves   mostly basal, narrowly to broadly oblanceolate (to 30 cm long), tapering to a clasping base; cauline leaves alternate, lanceolate, often villous-tomentose when young, becoming glabrate, black-glandular

flowers   heads (3–8), large and showy yellow-orange, loose corymbs, disk flowers (2.0–2.5 cm wide), rays narrow (1.5–2.5 cm long), petal 3-toothed at apex; phyllary bracts in 2 series

fruit   pubescent achenes (3 mm long), pappus of hyaline awn-pointed scales

other   stem villous to glabrous; veins in lower leaf appear parallel

HISTORIC, FOOD, AND MEDICINAL USES   none

LIVESTOCK LOSSES   poisonous to sheep, causes "spewing sickness" due to a glucoside; considerd poisonous to cattle, however since it is unpalatable, it is seldom grazed by cattle

FORAGE VALUE   poor, will be consumed by sheep and other animals when other forage is scarce

HABITAT   moist slopes and well-drained meadows

Bitterweed (*Hymenoxys odorata*)

268

Family	ASTERACEAE
Tribe	HELIANTHEAE
Species	*Hymenoxys odorata* DC.
Common Name	Bitterweed (bitter rubberweed, limonillo)
Life Span	Annual
Origin	Native
Season	Cool

GROWTH CHARACTERISTICS   forb (7–45 cm tall), much-branched, germinates in mid-winter to late winter, reproduces by seeds

DISTINGUISHING CHARACTERISTICS

leaves   alternate (3–6 cm long), deeply pinnatifid into 3–15 narrow-linear divisions, glandular-pubescent (appearing pitted), aromatic when crushed

flowers   many small heads (0.3–1.2 cm wide), corymbose; ray flowers yellow and horn-shaped; disk flowers yellow; outer phyllary bracts united near base, inner phyllary bracts convergent to an awn and green-tipped

fruit   achenes; pappus acuminate, awn-pointed

other   entire plant sparingly pubescent to glabrate

HISTORIC, FOOD, AND MEDICINAL USES   none

LIVESTOCK LOSSES   poisonous to sheep especially in winter, toxicity increases under water stress, poisoning is accumulative in sheep

FORAGE VALUE   poor to worthless for all classes of livestock and wildlife

HABITAT   waste places, overgrazed plains, and rangeland

Prairie coneflower (*Ratibida columnaris*)

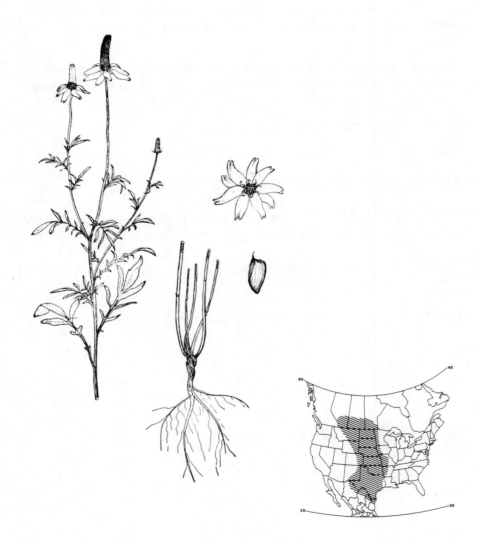

Family	ASTERACEAE
Tribe	HELIANTHEAE
Species	*Ratibida columnaris* (Sims) D. Don
Common Name	Prairie coneflower (upright prairie coneflower)
Life Span	Perennial
Origin	Native
Season	Warm

GROWTH CHARACTERISTICS   forb (0.2–1.2 m tall), branched, tap-root, flowers in all warm seasons with available moisture, reproduces by seeds

DISTINGUISHING CHARACTERISTICS

leaves   alternate (3–15 cm long including petiole), pinnately cleft to the midrib into 5–13 linear divisions, surfaces closely strigose-hirsute, margins mostly entire

flowers   heads borne singly at top of leafless branches; ray flowers yellow or red to brown at base of petals, drooping, oval in outline; disk flowers brown in a cylindric "column" (1.0–5.5 cm long)

fruit   compressed achenes (2 mm long), pappus of 2 tooth-like projections

other   occasional resinous dots may occur on foliage

HISTORIC, FOOD, AND MEDICINAL USES   Cheyenne Indians boiled leaves and stems to make a yellow solution applied externally to draw out poison of rattlesnake bites, also applied solution for relief from poison ivy; other Indian tribes made tea from the flowers and leaves

LIVESTOCK LOSSES   none

FORAGE VALUE   fair to good for sheep and wildlife, fair for cattle

HABITAT   prairies and plains

271

Mulesears (*Wyethia amplexicaulis*)

Family	ASTERACEAE
Tribe	HELIANTHEAE
Species	*Wyethia amplexicaulis* (Nutt.) Nutt.
Common Name	Mulesears (pe-ik)
Life Span	Perennial
Origin	Native
Season	Cool

GROWTH CHARACTERISTICS forb (30–60 cm tall), often grows in extensive patches, starts growth in March–April, flowers April–June, reproduces by seeds

DISTINGUISHING CHARACTERISTICS

leaves mostly basal, petiolate, elliptic or lanceolate-elliptic (20–60 cm long and 5–16 cm wide), cauline leaves reduced and sessile to clasping (mostly 9–25 cm long and 2–6 cm wide), glabrous, resinous

flowers heads several or sometimes solitary; rays bright yellow (2.5–5.0 cm long); disk flowers light yellow, receptacle chaffy; phyllary bracts herbaceous and broad, subequal

fruit achene (0.6–1.5 cm long), pappus a crown of scales

other stout taproot

HISTORIC, FOOD, AND MEDICINAL USES American Indians fermented roots for 2 days in a pit heated with hot stones to develop a sweet flavored food

LIVESTOCK LOSSES none

FORAGE VALUE flowers eaten by all classes of livestock, deer, and elk; sheep will graze young foliage

HABITAT open hillsides, dry meadows, and moist draws

Woolly mulesears (*Wyethia mollis*)

Family	ASTERACEAE
Tribe	HELIANTHEAE
Species	*Wyethia mollis* Gray
Common Name	Woolly mulesears (woolly wyethia)
Life Span	Perennial
Origin	Native
Season	Cool

GROWTH CHARACTERISTICS   forb (20–50 cm tall), tufted, often growing in dense stands, starts growth in March, flowers April–May, reproduces by seeds

DISTINGUISHING CHARACTERISTICS

leaves   mostly basal, lanceolate to oblong-ovate (20–40 cm long and 6–17 cm wide), petiolate, margin entire, cauline leaves reduced, densely tomentose especially when young, resin-dotted

flowers   heads (2–3 cm wide), 1–4 per plant; ray flowers deep yellow; disk flowers yellowish; 4–6 phyllary bracts, lanceolate, erect

fruit   achene (0.8–1.1 cm long), pubescent near apex; pappus a crown of rigid awns

other   woody taproot, stems tomentose

HISTORIC, FOOD, AND MEDICINAL USES   roots and fruits were eaten by American Indians

LIVESTOCK LOSSES   none

FORAGE VALUE   flowers readily eaten when young and tender; however, low in palatability for cattle and fair for sheep; grazed lightly by deer

HABITAT   dry wooded slopes, rocky openings, and grassy slopes

Threadleaf groundsel (*Senecio longilobus*)

Notes:

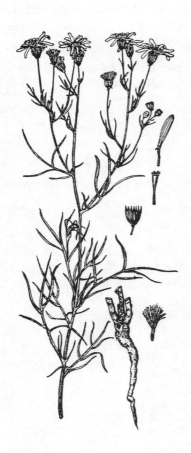

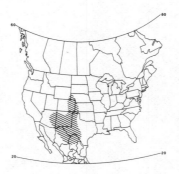

276

Family	ASTERACEAE
Tribe	SENECIONEAE
Species	*Senecio longilobus* Benth.
Common Name	Threadleaf groundsel
Life Span	Perennial
Origin	Native
Season	Warm

GROWTH CHARACTERISTICS   suffrutescent shrub (0.3–1.1 m tall), generally branched; plant leafy throughout, seeds germinate in early spring, flowers May–June, reproduces by seeds

DISTINGUISHING CHARACTERISTICS

    leaves   alternate, divided into 5–9 filiform divisions, divisions often unequal, margin entire and rolled inward, tomentose

    flowers   corymb, lemon yellow; small heads (0.9–1.6 cm high); phyllary bracts in one series, brown tipped

    fruit   achene, pubescent; pappus of capillary bristles

    other   stems tomentose

HISTORIC, FOOD, AND MEDICINAL USES   reportedly used by Indians for numerous medicinal purposes

LIVESTOCK LOSSES   poisonous to cattle, horses, and sheep (to a lesser extent); contains an alkaloid, most losses occur in late spring and summer

FORAGE VALUE   worthless to poor for livestock, sheep and goats may lightly browse it

HABITAT   dry plains, grasslands

Sawtooth butterweed (*Senecio serra*)

Family	ASTERACEAE
Tribe	SENECIONEAE
Species	*Senecio serra* Hook.
Common Name	Sawtooth butterweed (butterweed groundsel)
Life Span	Perennial
Origin	Native
Season	Warm

GROWTH CHARACTERISTICS   forb (0.6–1.2 m tall), reproduces by seeds and rootstalks, erect, branching only at inflorescence, starts growth in late spring, flowers July–August

DISTINGUISHING CHARACTERISTICS

leaves   alternate, lanceolate to linear (5–15 cm long), acute or acuminate at apex, margin sharply serrate, sessile or lower leaves petiolate, glabrous

flowers   in flat-topped groups, heads several (1.1–1.4 cm long and 5–8 mm wide), erect (not nodding); ray and disk flowers yellow; phyllary bracts in one series, thickened on back and near base, green with brown tips

fruit   achene, glabrous; pappus of capillary bristles

HISTORIC, FOOD, AND MEDICINAL USES   none

LIVESTOCK LOSSES   none

FORAGE VALUE   immature plants grazed by sheep, good for elk and deer, fair to poor for cattle

HABITAT   meadows, damp ground, and moist stream banks

Gray horsebrush (*Tetradymia canescens*)

Family	ASTERACEAE
Tribe	SENECIONEAE
Species	*Tetradymia canescens* DC.
Common Name	Gray horsebrush (spineless horsebrush)
Life Span	Perennial
Origin	Native
Season	Warm

GROWTH CHARACTERISTICS   shrub (0.2–1.0 m tall), deciduous, herbaceous above, woody below, stiffly branched, flowers June–September, reproduces by seeds

DISTINGUISHING CHARACTERISTICS

leaves   simple, alternate, woolly-canescent, linear to lanceolate (0.6–3.2 cm long), entire, prominent midrib

flowers   discoid; in terminal compact cymes, yellow to cream-colored; 4 phyllary bracts, margins thin and scarious, gray-tomentose

fruit   achene, rounded, silky-hairy

other   stems short and stout, silvery canescent-tomentose until maturity, becomes glabrous; nodes are prominent causing a "knobby" appearance

HISTORIC, FOOD, AND MEDICINAL USES   American Indians made a tonic from leaves and roots for uterine disorders

LIVESTOCK LOSSES   causes photosensitization in sheep, symptoms are called "big head" or "swell head" from swelling of the head and facial features; alkaloids may also cause liver damage in sheep

FORAGE VALUE   poor to worthless livestock browse; poisonous, resinous substance

HABITAT   barren plains, foothills, and plains on sandy or rocky soils

Tansymustard (*Descurainia pinnata*)

Family	BRASSICACEAE
Species	*Descurainia pinnata* (Walt.) Britt.
Common Name	Tansymustard
Life Span	Annual
Origin	Native
Season	Cool

GROWTH CHARACTERISTICS   forb (to 80 cm tall), branched or simple, one of the first spring annuals to appear, reproduces from seeds

DISTINGUISHING CHARACTERISTICS

leaves   alternate, deeply pinnately divided, lower leaves twice-divided (3–9 cm long) with segments obovate, upper leaves reduced in size with linear to oblanceolate segments, pubescent

flowers   racemose, 4 sepals (1.5–2.5 mm long) about equaling the 4 yellow petals, style minute

fruit   silique, narrowly club-shaped (0.5–2.0 cm long), usually curved upward (1.5–2.0 mm wide), seeds in 2 rows

other   taproot

HISTORIC, FOOD, AND MEDICINAL USES   seed pods used by Indians to make pinole flour, and young growth was used as a potherb

LIVESTOCK LOSSES   known to be poisonous to cattle in the Southwest, cattle get "paralyzed tongue" or an inability to swallow food and water; generally when this occurs, it is from consuming large quantities; however, this may not be easily avoided since this plant is green earlier in spring than most forage species

FORAGE VALUE   good when immature

HABITAT   waste places, prairies, dry or sandy soils, open woods

Desert princesplume (*Stanleya pinnata*)

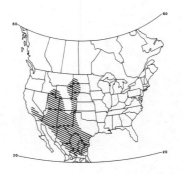

Family	BRASSICACEAE
Species	*Stanleya pinnata* (Pursh) Britt.
Common Name	Desert princesplume
Life Span	Perennial
Origin	Native
Season	Cool

GROWTH CHARACTERISTICS   forb (0.6–1.6 m tall), often suffrutescent, mostly unbranched except from the base, flowers April–May

DISTINGUISHING CHARACTERISTICS

leaves   alternate, upper usually entire (0.6–6.4 cm long), lower entire to pinnately divided, slightly pubescent to glabrous

flowers   pedicellate, yellow, calyx of 4 sepals turned downward with age, corolla of 4 golden yellow petals with brown claws, 6 long exserted stamens, flowers arranged in terminal spike-like raceme

fruit   silique, linear (5–10 cm long), on a stipe (1.0–2.5 cm long), seeds in 1 row

other   twigs light green and glaucous

HISTORIC, FOOD, AND MEDICINAL USES   pioneers and American Indians cooked stems and leaves (cabbage-like taste)

LIVESTOCK LOSSES   poisonous, accumulates selenium from the soil

FORAGE VALUE   worthless

HABITAT   dry hills, valleys, desert washes; reliable indicator of seleniferous soil

Snowberry (*Symphoricarpos albus*)

Family	CAPRIFOLIACEAE
Species	*Symphoricarpos albus* (L.) Blake
Common Name	Snowberry (common snowberry)
Life Span	Perennial
Origin	Native
Season	Cool

GROWTH CHARACTERISTICS   shrub (15–90 cm tall), thicket-forming, rhizomatous, flowers May–July, fruits August–September

DISTINGUISHING CHARACTERISTICS

leaves   simple, opposite, pedicellate, oval or ovate (0.9–3.8 cm long and 0.8–2.5 cm wide), margin entire (rarely wavy), upper side densely puberulent to glabrous, lower side pale (pubescent, especially on veins), sometimes appearing brown-speckled

flowers   arranged in spike-like racemes; calyx 5-toothed; corolla pink-campanulate (4–9 mm long), pedicellate

fruit   ovoid drupe (0.6–1.6 cm diameter), solitary or paired in leaf axils, white

other   winter buds with puberulent scales or often spreading curly pubescent; twigs yellow-brown covered with fine curled hairs especially at the nodes

HISTORIC, FOOD, AND MEDICINAL USES   American Indians made a tonic from the roots, an eyewash from the bark, and all parts were crushed and applied to wounds

LIVESTOCK LOSSES   poisonous saponin has been reported to be present in the leaves

FORAGE VALUE   sheep and goat winter browse; important wildlife food and cover for grouse, partridge, pheasant, and quail; occasionally browsed by mule deer

HABITAT   wooded hillsides, rocky slopes

Buckbrush (*Symphoricarpos orbiculatus*)

Notes:

Family	CAPRIFOLIACEAE
Species	*Symphoricarpos orbiculatus* Moench
Common Name	Buckbrush (coralberry, wolfberry)
Life Span	Perennial
Origin	Native
Season	Cool

GROWTH CHARACTERISTICS   shrub (1 m tall), rhizomatous, forming dense colonies, flowers April–May, fruits September, reproduces also by seeds

DISTINGUISHING CHARACTERISTICS

  leaves   simple (3–5 cm long and 1.5–3.5 cm wide), opposite, pedicellate, oval to ovate, margin entire sometimes undulate, upper surface dull green, glabrous or glabrate, veins impressed, lower surface paler and pubescent

  flowers   in short spikes in leaf axils, sessile; calyx 5-parted; corolla campanulate (3–4 mm long), green-white to purple

  fruit   red to black globose drupe (4.5–6.0 mm in diameter), calyx persistent, fruit in clusters

  other   buds gray-brown (0.5 mm long); twigs brown and flexible with white curved hairs

HISTORIC, FOOD, AND MEDICINAL USES   leaves were steeped by Blackfoot Indians to make a wash for sore eyes, berries were used as famine food, and boiled berries were given to horses as a diuretic

LIVESTOCK LOSSES   none

FORAGE VALUE   important wildlife browse; food and cover for several song and game birds; important for erosion control

HABITAT   hillsides, open woods, and riverbanks

Fourwing saltbush (*Atriplex canescens*)

Notes:

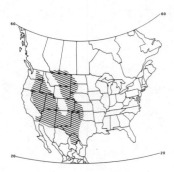

Family	CHENOPODIACEAE
Species	Atriplex canescens (Pursh) Nutt.
Common Name	Fourwing saltbush, (costilla de vaca, chamizo)
Life Span	Perennial
Origin	Native
Season	Evergreen

GROWTH CHARACTERISTICS   shrub (to 7.5 m tall), erect, several branches, light green, drought tolerant, roots penetrate soil to about 6 m, reproduces by seeds

DISTINGUISHING CHARACTERISTICS

leaves   alternate (0.9–5.0 cm long and 0.3–1.2 cm wide), sessile or nearly so, linear to oblong to spatulate, 1-nerved, thick, densely gray-scurfy, margins slightly inrolled

flowers   unisexual, plants dioecious; staminate heads in spicate terminal panicles; pistillate heads in axillary clusters and panicles with several leafy bracts

fruit   4-winged, varying in size and shape, wing margins entire to laciniate

other   twigs stout, gray-scurfy

HISTORIC, FOOD, AND MEDICINAL USES   Southwestern Indians ground seeds to make baking powder for bread; pollen will cause hay fever and is used in immunization extracts

LIVESTOCK LOSSES   concentrated feeding has reportedly caused scours in cattle, secondary or facultative selenium absorber

FORAGE VALUE   valuable browse for cattle, sheep, goats, and deer; crude protein level of about 7–9%

HABITAT   grassy uplands, sandy deserts, and alkali flats

Shadscale saltbush (*Altriplex confertifolia*)

Family	CHENOPODIACEAE
Species	*Atriplex confertifolia* (Torr. & Frem.) Wats.
Common Name	Shadscale saltbush (saladillo)
Life Span	Perennial
Origin	Native
Season	Cool

GROWTH CHARACTERISTICS   shrub (15–90 cm tall), forming round-ed clumps with numerous erect spiny branches, starts growth in late spring, seed matures in October

DISTINGUISHING CHARACTERISTICS

leaves   simple (0.9–1.9 cm long and 0.3–1.2 cm wide), alternate, orbicu-lar to ovate, obovate or elliptic, surfaces gray-scurfy, short petiolate; crowded at first, but later deciduous exposing spiny twigs

flowers   unisexual, plants dioecious; staminate spikes with leafy bracts; pistillate flowers few or solitary in upper leaf axils

fruit   achenes, bracts sessile, united around fruit, free and divergent above

other   twigs stout, light brown and scurfy, turning gray and glabrous with age, ending in spines

HISTORIC, FOOD, AND MEDICINAL USES   American Indians ground the fruits into flour

LIVESTOCK LOSSES   none

FORAGE VALUE   used by all classes of livestock, especially as winter and spring browse; seeds provide food for game and songbirds

HABITAT   alkaline desert valleys, hills, and bluffs

Saltbush (*Atriplex gardneri*)

294

Family	CHENOPODIACEAE
Species	*Atriplex gardneri* (Moq.) D. Dietr.
Common Name	Saltbush (Gardner saltbush, saladillo)
Life Span	Perennial
Origin	Native
Season	Evergreen

GROWTH CHARACTERISTICS   shrub (20–50 cm tall), woody at the base, much-branched

DISTINGUISHING CHARACTERISTICS

leaves   alternate (lower leaves may be opposite), oblong-linear to obovate, base narrowed, apex rounded (2.5–5.0 cm long and 0.3–1.0 cm wide), margin entire, gray-green, scurfy

flowers   unisexual, plants dioecious; in axillary groups or in leafy paniculate spikes

fruit   2-winged, bracts sessile or short-petiolate, lanceolate to lance-ovate, bracts united to form a subfusiform fruit (5–8 mm long), margin entire to slightly toothed, surface tubercled or crested

other   stem is tomentose, often very white

HISTORIC, FOOD, AND MEDICINAL USES   American Indians ground parched seeds to make pinole flour

LIVESTOCK LOSSES   none

FORAGE VALUE   important winter browse on western ranges for cattle and sheep

HABITAT   alkaline prairie soils

Winterfat (*Ceratoides lanata*)

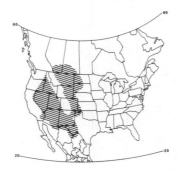

Family	CHENOPODIACEAE
Species	*Ceratoides lanata* (Pursh) J. T. Howell
Common Name	Winterfat (lambstail, white sage)
Life Span	Perennial
Origin	Native
Season	Cool

GROWTH CHARACTERISTICS suffrutescent shrub (20–80 cm tall), woody at base, branches stout and erect, flowers April–May, fruit matures July–August, reproduces by seeds, resistant to drought

DISTINGUISHING CHARACTERISTICS

leaves alternate, or somewhat fascicled, sessile to short-petioled, linear to narrowly lanceolate (1.3–4.4 cm long), margins entire and enrolled, distinct protruding midrib, densely pubescent with simple or stellate hairs

flowers unisexual, plants monoecious; in dense axillary clusters; calyx 4-lobed distinctly hairy; styles 2; pubescent

fruit urticle enclosed by 2 bracts, 2-horned above with tufts of white hair

other twigs are stout and densely hairy, branches may become spinescent

HISTORIC, FOOD, AND MEDICINAL USES American Indians treated fevers with a decoction from the leaves, Blackfoot Indians soaked the leaves in warm water to make a hair wash

LIVESTOCK LOSSES none

FORAGE VALUE valuable browse, succulent through winter for livestock browse for elk, mule deer, and rabbits, high in crude protein; valuable erosion control

HABITAT stony hillsides, dry soils of mesas and plains

Spiny hopsage (*Grayia spinosa*)

Family	CHENOPODIACEAE
Species	*Grayia spinosa* (Hook.) Moq.
Common Name	Spiny hopsage (Grays saltbush, spiny sage)
Life Span	Perennial
Origin	Native
Season	Warm

GROWTH CHARACTERISTICS   shrub (25–40 cm tall), deciduous, stiff and much-branched

DISTINGUISHING CHARACTERISTICS

leaves   alternate, sessile, often in fascicles of 2, oblanceolate to oblong-lanceolate (1–3 cm long), margins entire, gray-green, scurfy, fleshy

flowers   unisexual, plants dioecious; male clusters axillary in dense terminal spikes; female flowers in terminal racemose spikes

fruit   2-winged, formed from bracts, membranous (0.6–1.2 cm long), thin, margins entire, often red-tinged

other   young branches light maroon in color, older branches with stringy exfoliating bark, stems have distinct white striations, twigs are sharp or spine-like at tips

HISTORIC, FOOD, AND MEDICINAL USES   American Indians ground parched seeds to make pinole flour

LIVESTOCK LOSSES   spines may cause minor injury

FORAGE VALUE   browsed in fall, winter, and spring by all classes of livestock; seeds valuable for fattening sheep

HABITAT   mesas and flats; alkaline soils, limestones, gravelly, and dry heavy clay soils

Halogeton (*Halogeton glomeratus*)

Family	CHENOPODIACEAE
Species	*Halogeton glomeratus* (Bieb.) C. A. Mey
Common Name	Halogeton
Life Span	Annual
Origin	Introduced (from Asia)
Season	Warm

GROWTH CHARACTERISTICS

forb (5–15 cm tall), much-branched from the base, germinates February–May, seeds mature by October, prolific seed producer

DISTINGUISHING CHARACTERISTICS

leaves   alternate (0.6–2.0 cm long), round in cross-section, apex truncate with a bristle-like tip

flowers   in compact clusters in leaf axils, 2-bracted, greenish-yellow, numerous; 2 kinds, longer flowers with wing-tipped sepals surrounding seed cases, smaller flowers with tooth-like sepals at the apex

fruit   compressed seed (1 mm long), wide-winged bracts (sepals) persistent on black seed, some seed light tan-colored

other   glaucous; taproot; fleshy, bluish-green in early summer, becoming yellow or red in late summer

HISTORIC, FOOD, AND MEDICINAL USES   none

LIVESTOCK LOSSES

poisonous especially to sheep, contains toxic amounts of sodium, potassium, and calcium oxalates; first signs of poisoning occur 2–6 hours after an animal eats a fatal amount, and death occurs in 9 to 11 hours

FORAGE VALUE

palatable to both cattle and sheep; provides usable forage when mixed in small quantities with other forage plants

HABITAT   alkaline and disturbed areas

Greenmolly summercypress (*Kochia americana*)

Notes:

Family	CHENOPODIACEAE
Species	*Kochia americana* Wats.
Common Name	Greenmolly summercypress (perennial summercypress, red sage)
Life Span	Perennial
Origin	Native
Season	Warm

GROWTH CHARACTERISTICS   suffrutescent shrub (10–30 cm tall), from a woody much-branched crown

DISTINGUISHING CHARACTERISTICS

leaves   alternate or often fascicled, sessile, numerous, linear (0.6–2.5 cm long), terete to flat when dried, fleshy, erect or ascending, surfaces silky-pubescent to glabrate

flowers   perfect, axillary, solitary or in groups of 2–3, white, tomentose; calyx 5-parted, persistent wrapping around the fruit

fruit   urticle, depressed globose (2 mm long); calyx forming 5 horizontal wings

other   stems white; dried plants may appear black

HISTORIC, FOOD, AND MEDICINAL USES   none

LIVESTOCK LOSSES   may accumulate nitrates

FORAGE VALUE   excellent for sheep, cattle, and deer; high in protein during the fall; deficient in phosphorus throughout the year

HABITAT   alkaline flats of cold deserts

Kochia (*Kochia scoparia*)

Family	CHENOPODIACEAE
Species	*Kochia scoparia* (L.) Schrad.
Common Name	Kochia (fireweed, summercypress, belvedere)
Life Span	Annual
Origin	Introduced (from Eurasia)
Season	Warm

GROWTH CHARACTERISTICS   forb, erect, much-branched, (0.5–1.5 m tall), grows from early summer to fall, plant size depends on available moisture

DISTINGUISHING CHARACTERISTICS

leaves   alternate, linear (2–7 cm long and 3–8 mm wide), prominently 3- to 5-veined, taper at base to a slender petiole, pubescent

flowers   sessile in axils of reduced leaf-like bracts, calyx winged (1.5–2.0 mm wide)

fruit   seed (1.5 mm in diameter), 5-cleft calyx persistent around fruit

other   stems often become red with age

HISTORIC, FOOD, AND MEDICINAL USES   escaped ornamental, may cause respiratory problems

LIVESTOCK LOSSES   may cause nitrate poisoning, may cause photosensitization

FORAGE VALUE   generally considered a weed, however is preferred by livestock and shows high productivity and nutritive value comparable to alfalfa

HABITAT   wastelands, disturbed sites, and fields

Russian thistle (*Salsola iberica*)

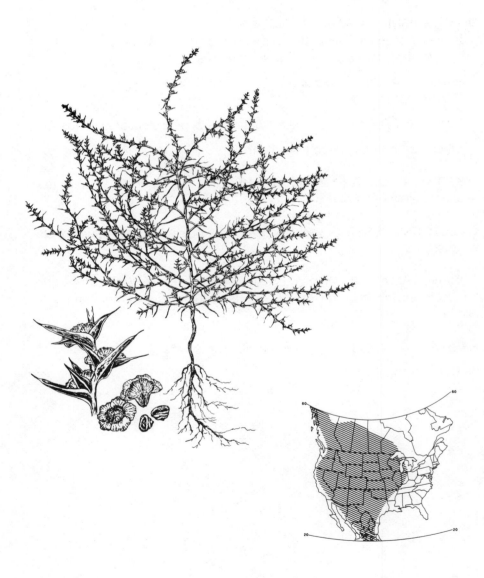

Family	CHENOPODIACEAE
Species	*Salsola iberica* Sennen & Pau
Common Name	Russian thistle (tumbleweed, rodadora)
Life Span	Annual
Origin	Introduced (from Eurasia)
Season	Warm

GROWTH CHARACTERISTICS   forb (30—80 cm tall), much-branched, rounded in shape, usually green and purple-striped or red, germinates March—April, flowers July—August, reproduces by seeds

DISTINGUISHING CHARACTERISTICS

leaves   alternate, linear to filiform (somewhat rounded) (1.2—3.0 cm long), sessile to clasping, pointed at apex, upper leaves are thickened at base and generally enclose fruit

flowers   perfect, 5-merous, small, sessile, in leaf axils, subtended by 2 bracts and a leaf

fruit   pericarp, fleshy; calyx of 5 persistent sepals, membranous white or pink (3—8 mm in diameter together with the fruit), seed rounded, black and shiny

other   plants often tinged with red or purple, glabrous to pubescent; seedlings have linear, thread-like leaves appearing opposite; stems are red below cotyledons, striped with white above

HISTORIC, FOOD, AND MEDICINAL USES   young shoots can be used as a potherb; seeds can be ground into meal; first introduced into North Dakota in flax seed

LIVESTOCK LOSSES   mechanical injury from sharp-pointed leaves, may accumulate nitrates, may contain oxalates

FORAGE VALUE   fair in early spring, but when plant matures the leaves become sharp-pointed and the plant is worthless; hay can be made from young plants

HABITAT   nearly all soil types of waste places and disturbed sites

Black greasewood (*Sarcobatus vermiculatus*)

Family	CHENOPODIACEAE
Species	*Sarcobatus vermiculatus* (Hook.) Torr.
Common Name	Black greasewood (greasewood, chico)
Life Span	Perennial
Origin	Native
Season	Warm

GROWTH CHARACTERISTICS   shrub (1.0–2.5 m tall), rounded, erect or spreading, leaves appear in late spring, flowers mid-summer, seeds mature by September

DISTINGUISHING CHARACTERISTICS

leaves   simple (1–4 cm long), alternate, erect, linear to linear-filiform, fleshy, margins entire, glabrous or with some stellate pubescence, olive-green

flowers   unisexual, plants monoecious or sometimes dioecious; staminate flowers in terminal ament-like spikes (0.7–3.0 cm long); pistillate flowers solitary or paired in leaf axils, often tinged with red

fruit   coriaceous (4–5 mm long and 2.5–3.5 mm wide), winged at middle

other   stems smooth with white exfoliating bark, becoming gray with maturity; thorns and branches diverge at 90° angles

HISTORIC, FOOD, AND MEDICINAL USES   wood is sometimes used for fuel; American Indians used sticks as planting tools

LIVESTOCK LOSSES   soluble oxalates have caused mass mortality in flocks of sheep, cattle are rarely poisoned

FORAGE VALUE   valuable browse in winter for domestic livestock, important wildlife food for porcupines, jackrabbits, prairie dogs, and chipmunks

HABITAT   flat ground, barren or alkaline soils

One-seed juniper (*Juniperus monosperma*)

Family	CUPRESSACEAE
Species	*Juniperus monosperma* (Engelm.) Sarg.
Common Name	One-seed juniper (cherrystone juniper, tascate, enebro)
Life Span	Perennial
Origin	Native
Season	Evergreen

GROWTH CHARACTERISTICS   shrub or small tree (to 12 m tall), with a rounded crown, branching near the ground, often with several trunks, flowers March–April, fruits September

DISTINGUISHING CHARACTERISTICS

leaves   opposite or in 3's, overlapping and appressed against branches, scale or awl-like, spiny on older branches, triangular shape (1.0–3.5 mm long), taper to a sharp point, generally glandular on back, gray-green

flowers   unisexual, plants dioecious; male cones small and cylinder-shaped, female cones small and globe-shaped

fruit   fleshy, berry-like cones (5–7 mm in diameter), blue to copper-colored; 1 seed per fruit

other   bark thin, ash-gray to gray

HISTORIC, FOOD, AND MEDICINAL USES   fruits were gathered and eaten by American Indians; wood used by Indians for prayer sticks, war bows, and instruments; Indians obtained green dye from bark; ground fruit to make flour and bread; fibrous bark was used for mats, saddles, and breech-cloths

LIVESTOCK LOSSES   may cause abortion

FORAGE VALUE   occasionally browsed by goats; fruits are eaten by deer, quail, foxes, chipmunks, squirrels, songbirds, and coyotes

HABITAT   hillsides or rocky exposed slopes and ledges

Rocky Mountain juniper (*Juniperus scopulorum*)

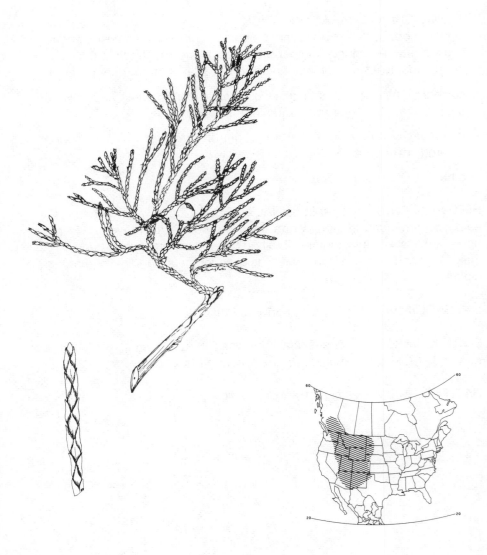

Family	CUPRESSACEAE
Species	*Juniperus scopulorum* Sarg.
Common Name	Rocky Mountain juniper (Rocky Mountain cedar)
Life Span	Perennial
Origin	Native
Season	Evergreen

GROWTH CHARACTERISTICS   shrub or small tree (to 12 m), crown dome-like, irregular, branching near the ground, flowers in April, fruit ripens November and December of the second year following pollination

DISTINGUISHING CHARACTERISTICS

leaves   opposite or in whorls of 3, closely appressed but not overlapping, ovate to ovate-elliptic (mostly 2 mm long), margins smooth, usually with an obscure elongate-elliptic gland on under surface, juvenile leaves awl-shaped and sharply pointed

flowers   unisexual, plants dioecious, cones borne at tips of branches male cone oblong (2 mm long), female cones globe-shaped, scales spreading

fruit   fleshy, berry-like cones (5–8 mm in diameter), bright blue, glaucous, resinous-juicy, seeds mostly 2

HISTORIC, FOOD, AND MEDICINAL USES   American Indians ate fruit raw or cooked, used as flavorings for meat and gin, fruit and young shoots were boiled for tea, berries were ground for mush and cakes, wax from berries used in candles; currently used as ornamental, in shelterbelts, for fence posts

LIVESTOCK LOSSES   may cause abortions

FORAGE VALUE   generally not consumed by livestock, important browse plant for antelope, mule deer, and bighorn sheep; birds utilize fruits

HABITAT   dry, rocky hillsides; often on undeveloped, erodable soils; grows best on calcareous and somewhat alkaline soils

Threadleaf sedge (*Carex filifolia*)

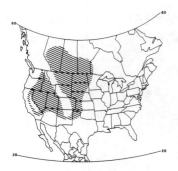

Family	CYPERACEAE
Species	*Carex filifolia* Nutt.
Common Name	Threadleaf sedge (blackroot)
Life Span	Perennial
Origin	Native
Season	Cool

GROWTH CHARACTERISTICS   grass-like plant (8–30 cm tall), densely caespitose

DISTINGUISHING CHARACTERISTICS

leaves   grass-like, linear with parallel veins, 2–3 per culm, near base; blades rolled (3–20 cm long; 0.25 mm wide), stiff, light green; sheaths brown near base; leaves and sheaths glabrous

flowers   in solitary spikes (1–3 cm long and 3–5 mm wide); pistillate below and staminate above; scales obtuse, light reddish-brown with white margins; perigynia triangular (3.0–3.5 mm long), straw-colored; beak truncate, hyaline; stigmas 3, black

fruit   achenes (2.3–3.0 mm long), triangular

other   roots are fibrous, stout, and black

HISTORIC, FOOD, AND MEDICINAL USES   American Indians used culm bases as famine food

LIVESTOCK LOSSES   none

FORAGE VALUE   excellent for livestock and wildlife, extremely valuable early spring growth, maintains high palatability

HABITAT   dry plains and ridges

Elk sedge (*Carex geyeri*)

Notes:

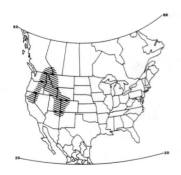

316

Family	CYPERACEAE
Species	*Carex geyeri* Boott
Common Name	Elk sedge (Geyer sedge, pine sedge)
Life Span	Perennial
Origin	Native
Season	Cool

GROWTH CHARACTERISTICS   grass-like plant (10–40 cm tall), clustered or loosely caespitose; reproduces by seeds and by thick, scaly rhizomes

DISTINGUISHING CHARACTERISTICS

leaves   grass-like (linear with parallel veins); blades about same length as culm (2.0–3.5 mm wide), flat or channeled, light green; sheaths tight, hyaline on upper surface

flowers   in solitary spikes (0.5–2.5 cm long and 1.5–3.0 mm wide), pistillate below and staminate above; scales wider and longer than perigynia, ovate, lower ones short-awned with broad hyaline margins; perigynia, triangular (6 mm long and 2.5 mm wide), greenish-straw or brown; short beak; stigmas 3

fruit   achenes

other   vegetative reproduction by creeping rhizomes, stem triangular in cross section

HISTORIC, FOOD, AND MEDICINAL USES   American Indians boiled and ate the culm bases

LIVESTOCK LOSSES   none

FORAGE VALUE   good for cattle and elk, fair for sheep and deer; it will be grazed in early spring since it greens up earlier than most forages

HABITAT   dry, open slopes and open woods

Russet buffaloberry (*Shepherdia canadensis*)

Notes:

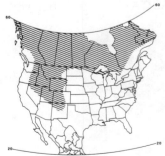

Family	ELAEAGNACEAE
Species	*Shepherdia canadensis* (L.) Nutt.
Common Name	Russet buffaloberry (Canadian buffaloberry, rabbitberry)
Life Span	Perennial
Origin	Native
Season	Cool

GROWTH CHARACTERISTICS   shrub (1–3 m tall), spreading, flowers April–June, fruit matures July–September, flowers appear before leaves

DISTINGUISHING CHARACTERISTICS

leaves   opposite, petiolate, oval to egg-shaped, (1–6 cm long), apex obtuse, base rounded, margin entire, dull green on upper surface, lower surface silvery and rusty scurfy-spotted

flowers   small, unisexual, plants dioecious, yellow on inside, brown on outside, solitary or in clusters at ends of branches; staminate with 4-lobed united calyx, 8 stamens, no petals; pistillate flowers with united urn-shaped 4-lobed calyx enclosing the ovary, no petals

fruit   berry-like (4–6 mm long), roundish, yellow or red, juicy

other   branches brown-scurfy, opposite

HISTORIC, FOOD, AND MEDICINAL USES   fruits are edible, cooked or raw, although insipid; fruits may be whipped with sugar for a dessert; commonly causes diarrhea

LIVESTOCK LOSSES   none

FORAGE VALUE   fair for sheep before frost, otherwise considered worthless, seldom browsed by livestock or game animals; fruits are consumed by many kinds of birds

HABITAT   limestone slopes and ledges or moist wooded slopes

Longleaf Mormon tea (*Ephedra trifurca*)

Family	EPHEDRACEAE
Species	*Ephedra trifurca* Torr.
Common Name	Longleaf Mormon tea (joint fir, ephedra, cañutillo)
Life Span	Perennial
Origin	Native
Season	Evergreen

GROWTH CHARACTERISTICS   shrub (to 2 m tall), erect, branches terete and rigid

DISTINGUISHING CHARACTERISTICS

leaves   scale-like, whorled in groups of 3 (0.5–1.3 cm long), united for ½ to ¾ of their total length, persistent and spinose on older branches, long internodes (3–9 cm long), sheath membranous, becoming fibrous with age

flowers   unisexual, plants dioecious; staminate cones subtended by 2 bracts; pistillate cones of 2 erect ovules enclosed in an urn-shaped involucre of scales

fruit   cones, solitary or numerous at upper nodes, elliptic to obovate, membranous-margined, reddish-brown

other   pale yellow-green, spinosely-tipped round branches, branches are photosynthetic, portions of stems are inflated

HISTORIC, FOOD, AND MEDICINAL USES   American Indians and Mexicans have long used stem decoctions as a cooling beverage and have eaten seeds; common name Mormon tea is derived from use as a beverage for Latter-Day-Saint pioneers in the American West; contains the drug ephedrine

LIVESTOCK LOSSES   none

FORAGE VALUE   poor for livestock, however, is heavily grazed in emergency situations; browsed by bighorns and jackrabbits; seeds are eaten by scaled quail

HABITAT   semi-desert foothills, dry creek beds, coarse soils

Pointleaf manzanita (*Arctostaphylos pungens*)

Family	ERICACEAE
Species	*Arctostaphylos pungens* H.B.K.
Common Name	Pointleaf manzanita (Mexican manzanita)
Life Span	Perennial
Origin	Native
Season	Evergreen

GROWTH CHARACTERISTICS   shrub (2–3 m tall), branched from base forming thickets, root sprouter, flowers January–March, fruit matures April–July

DISTINGUISHING CHARACTERISTICS

leaves   alternate (1.5–3.0 cm long), leathery, petiolate, oblong to oblanceolate, margin entire, bright green on both sides, apex acute

flowers   in racemes, white or pink; calyx 5-lobed and persistent; petals urceolate (6 mm long), 5-lobed at top; 10 stamens; superior ovary

fruit   drupe, berry-like, dark brown to black, glabrous, several nutlets which are ridged on back

other   stems rigid, white-tomentose on new growth, later becoming red and smooth

HISTORIC, FOOD, AND MEDICINAL USES   fruit is sold in Mexican markets and used for jellies; Mexicans use leaves and fruits as household remedies for dropsy, bronchitis, and venereal diseases

LIVESTOCK LOSSES   none

FORAGE VALUE   goats graze leaves and browse young twigs, in spring goats will peel back bark presumably for sap, fruit eaten by grouse, skunks, deer, quail, bears, and coyotes

HABITAT   rocky mesas, dry slopes, and mountain slopes

Bearberry (*Arctostaphylos uva-ursi*)

Family	ERICACEAE
Species	*Arctostaphylos uva-ursi* (L.) Spreng.
Common Name	Bearberry (manzanita, kinnikinnick)
Life Span	Perennial
Origin	Native
Season	Evergreen

GROWTH CHARACTERISTICS   shrub, prostrate or at least depressed, much-branched, terminal branches erect (15–20 cm long), flowers April–July, fruit matures July–October

DISTINGUISHING CHARACTERISTICS

leaves   alternate petiolate, obovate to oblong or oval (1.3–3.2 cm long and 0.5–1.0 cm wide), narrowed to the base, apex obtuse or rounded, margin entire, upper surface bright green and lustrous, lower surface pale

flowers   in racemes, white or pink, small and nodding; corolla urceolate with 5 short and recurved lobes; stamens 10; 5 short sepals

fruit   drupe, berry-like, bright red and lustrous (6–8 mm in diameter); nutlets 5, ridged

other   bark thin, becoming shreddy with age

HISTORIC, FOOD, AND MEDICINAL USES   fruit is edible if cooked, used by early settlers for treating urinary disorders; tobacco substitute

LIVESTOCK LOSSES   none

FORAGE VALUE   little value for livestock, browsed by mountain sheep, black-tailed and white-tailed deer; fruit provides food for grouse and wild turkeys

HABITAT   dry, well-drained soils, sparsely wooded areas or slightly above timberline

Guajillo (*Acacia berlandieri*)

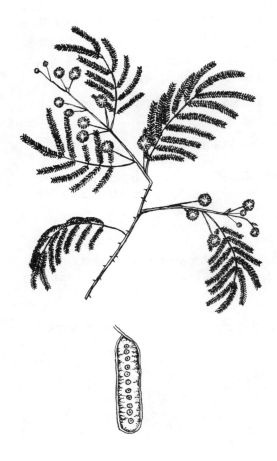

Family	FABACEAE
Species	*Acacia berlandieri* Benth.
Common Name	Guajillo (Berlandier acacia)
Life Span	Perennial
Origin	Native
Season	Warm

GROWTH CHARACTERISTICS   shrub (1–4 m tall), deciduous, several main stems from the base, flowers November–March, fruit matures June–July, reproduces by seeds

DISTINGUISHING CHARACTERISTICS

leaves   bipinnately compound; leaflets long (4 mm) and narrow, delicate, almost fern-like in appearance, 30–50 pair, pubescent to nearly glabrous

flowers   white to yellow in axillary globose heads (1 cm in diameter); corolla 5-parted and pubescent; stamens numerous and exserted

fruit   legume, large and flat (8–15 cm long and 1.5–2.5 cm wide), straight to somewhat curved (one somewhat straighter than the other), margins thickened dark brown, velvety-tomentose when mature

other   relatively thick and striate internodes generally armed with scattered prickles (1–3 mm long)

HISTORIC, FOOD, AND MEDICINAL USES   important honey plant, gums and dyes have been extracted from this shrub

LIVESTOCK LOSSES   will cause hydrocyanic acid poisoning in livestock when extremely large amounts are consumed

FORAGE VALUE   fair for wildlife and livestock

HABITAT   limestone ridges and caliche hills, sandy soils

Huisache (*Acacia farnesiana*)

Notes:

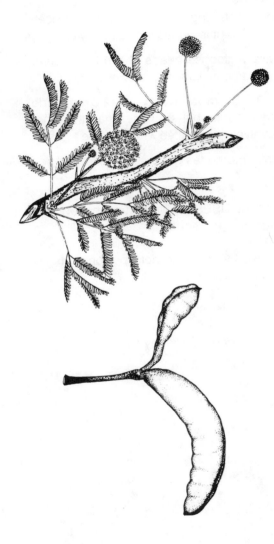

328

Family	FABACEAE
Species	*Acacia farnesiana* (L.) Willd.
Common Name	Huisache (sweet acacia)
Life Span	Perennial
Origin	Native
Season	Warm

GROWTH CHARACTERISTICS   shrub or small tree (2–10 m tall), deciduous, often with several trunks flaring upward, densely branched, flowers in early spring

DISTINGUISHING CHARACTERISTICS

leaves   bipinnately compound, 10–25 pairs of leaflets (3–5 mm long), linear, gray-green, unequal base

flowers   yellow-globose (1 cm in diameter), fragrant; corolla funnel-form and 5-lobed; numerous yellow exserted stamens

fruit   legume (2–8 cm long), tapered to both ends, nearly round in cross-section, black seeds in 2 rows, woody and stout, straight or curved

other   stipular paired spines, straight, white and rigid

HISTORIC, FOOD, AND MEDICINAL USES   formerly source of oils for "French" perfumes; wood has been used for fence posts, farm tools, and smaller wooden items; gummy sap can be used for manufacturing mucilage; ornamental; honey plant

LIVESTOCK LOSSES   spines may injure soft tissue

FORAGE VALUE   valuable food and cover for wildlife, poor for livestock

HABITAT   dry sandy soils in pinelands, hammocks, and disturbed areas

Catclaw acacia (*Acacia greggii*)

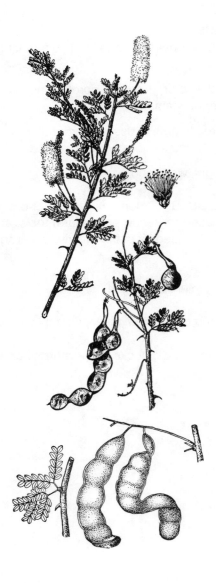

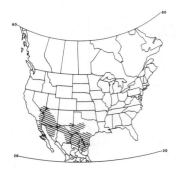

330

Family	FABACEAE
Species	*Acacia greggii* Gray
Common Name	Catclaw acacia (Gregg acacia, uña de gato)
Life Span	Perennial
Origin	Native
Season	Warm

GROWTH CHARACTERISTICS   shrub or small tree (to 6 m tall), deciduous, rounded and much-branched, prickles, thicket forming, flowers April–October, reproduces by seeds

DISTINGUISHING CHARACTERISTICS

leaves   bipinnately compound; 3–7 pairs of leaflets (3–6 mm long), obovate to narrowly oblong, base unequally contracted to a short petiole, light reticulate-veined, pubescent

flowers   creamy white in spikes (1 cm wide and 2- to 6-times longer); corolla of 5 petals; stamens numerous and exserted

fruit   legume (5–8 cm long and 1.5–2.0 cm broad), curved or often curled and contorted, constricted between the seeds, margins thickened, light brown or red

other   recurved prickles (catclaws) on internodes

HISTORIC, FOOD, AND MEDICINAL USES   wood used for fuel; Pima and Papago Indians made pinole flour from the pods and prepared as a mush; important honey plant; host plant for several lac insects which produce lac, a material used in varnish and shellac

LIVESTOCK LOSSES   spines may cause injury

FORAGE VALUE   food and cover for wildlife, main food source for some species of quail

HABITAT   dry valleys and ravines, sandy or gravelly arid mesas, washes, and canyon slopes

Blackbrush acacia (*Acacia rigidula*)

Notes:

Family	FABACEAE
Species	*Acacia rigidula* Benth.
Common Name	Blackbrush acacia (chaparro prieto)
Life Span	Perennial
Origin	Native
Season	Warm

GROWTH CHARACTERISTICS   shrub (1–3 m tall), several branches from base, often thicket-forming, flowers April–May, reproduces by seeds

DISTINGUISHING CHARACTERISTICS

leaves   bipinnately compound; leaflets few and large (0.6–1.5 cm long), 2–4 pairs, green, glabrous, oblong, oblique, apex rounded and mucronate, sometimes notched, base asymmetric, nerves conspicuous

flowers   white, sessile in axillary spikes, numerous, fragrant; calyx of 4–5 sepals; corolla white-yellow of 4–5 petals; several exserted stamens

fruit   legume (6–8 cm long), narrow, about as thick as broad, slightly constricted between seeds, puberulent, acuminate at apex

other   stem with paired, straight, long, rigid white spines at nodes; light to dark gray bark

HISTORIC, FOOD, AND MEDICINAL USES   sometimes used as an ornamental, flowers source for honey

LIVESTOCK LOSSES   spines may cause injury

FORAGE VALUE   fair for wildlife, poor for livestock

HABITAT   sandy and limestone areas, on ridge tops and plains; provides erosion control

Leadplant (*Amorpha canescens*)

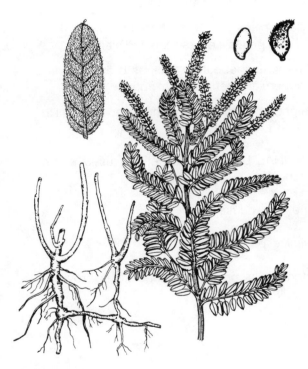

Family	FABACEAE
Species	*Amorpha canescens* (Nutt.) Pursh
Common Name	Leadplant
Life Span	Perennial
Origin	Native
Season	Warm

GROWTH CHARACTERISTICS　　suffrutescent shrub (to 1 m tall), branched taproot, reproduces from seeds, provides good soil cover

DISTINGUISHING CHARACTERISTICS

leaves　odd-pinnately compound, 7–24 pairs leaflets, crowded to overlapping, (0.9–1.8 cm long), rounded at base, acute to obtuse at apex, usually mucronate, gray-canescent, stipules linear (bristle-like)

flowers　in crowded racemes, rachis densely villous; calyx tube canescent (5 mm long); corolla of 1 petal, light blue to purplish; stamens 10, exserted

fruit　legume (4 mm long), densely tomentose-canescent

other　stems erect or ascending; whole plant has a gray-green or lead-colored appearance

HISTORIC, FOOD, AND MEDICINAL USES　　cultivated ornamental; American Indians smoked dried leaves and made tea from leaves

LIVESTOCK LOSSES　　none

FORAGE VALUE　　excellent, highly nutritive, and palatable for livestock and wildlife; grazed only in spring in some areas

HABITAT　　dry plains, hills, and prairies

Woolly loco (*Astragalus mollissimus*)

Notes:

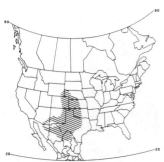

Family	FABACEAE
Species	*Astragalus mollissimus* Torr.
Common Name	Woolly loco (poisonvetch, nierba loco)
Life Span	Perennial
Origin	Native
Season	Cool

GROWTH CHARACTERISTICS   forb (10–35 cm tall), tufted, often robust and leafy, starts growth in March, fruit matures by June, reproduces by seeds

DISTINGUISHING CHARACTERISTICS

leaves   odd-pinnately compound, 11–35 leaflets (0.5–2.5 cm long and 0.2–1.5 cm wide), egg-shaped, rounded at apex, villous-tomentose, stipules free from base of leafstalks

flowers   in a raceme; calyx tube silky (7–9 mm long); corolla purple (1.8 cm long), keel petal rounded

fruit   legume, spreading or ascending, linear oblong (0.9–2.5 cm long), apex pointed, abruptly curved upwards

other   stems and foliage densely canescent throughout, hairs attached basally

HISTORIC, FOOD, AND MEDICINAL USES   famous in the history of the West as one of the causes of "locoed" horses

LIVESTOCK LOSSES   poisonous, can cause loco disease in horses, cattle, sheep, and goats; caused by lococine and selenium which is absorbed and accumulated by the plant during dry periods; the name loco disease is derived from the symptoms shown by animals after eating this plant

FORAGE VALUE   poor to worthless, animals may utilize woolly loco when other forage is not available

HABITAT   plains and prairies

Tailcup lupine (*Lupinus caudatus*)

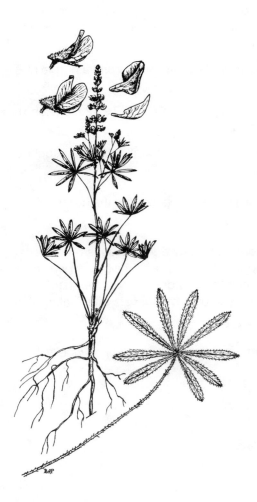

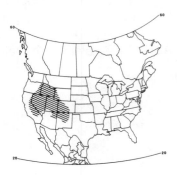

Family	FABACEAE
Species	*Lupinus caudatus* Kell.
Common Name	Tailcup lupine (spurred lupine)
Life Span	Perennial
Origin	Native
Season	Cool

GROWTH CHARACTERISTICS   forb (to 50 cm tall), flowers June—July, fruit matures July—August, reproduces by seeds

DISTINGUISHING CHARACTERISTICS

leaves   alternate, palmately compound into 7—9 linear to lance-shaped leaflets, stiff hairy to silky on both surfaces, silvery appearance, long petiolate

flowers   irregular, in racemes; calyx tube 3—4 mm long, spurred at the base; corolla (1.0—1.2 cm long), blue to deep violet; banner petal reflexed

fruit   legume

other   stems pubescent

HISTORIC, FOOD, AND MEDICINAL USES   an antiarrhythmic drug has been extracted for management of cardiac arrhythmias

LIVESTOCK LOSSES   poisonous, especially to sheep and horses, causing weakness and muscular trembling, alkaloids are concentrated in the seeds and occasionally in the young plants

FORAGE VALUE   poor for cattle and fair for sheep before the pods develop

HABITAT   dry, well-drained soils, slopes

Burclover (*Medicago hispida*)

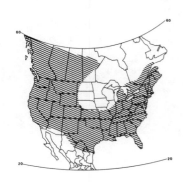

Family	FABACEAE
Species	*Medicago hispida* Gaertn.
Common Name	Burclover (toothed burclover, medic)
Life Span	Annual
Origin	Introduced (from Mediterranean region)
Season	Cool

GROWTH CHARACTERISTICS   forb, prostrate to ascendent (stems 20–50 cm long), flowers in early spring, matures in about 8 weeks

DISTINGUISHING CHARACTERISTICS

leaves   pinnately trifoliate (1.0–1.5 cm long), petiolate, leaflets cuneate-obovate, sharply toothed, leaflets sometimes with white or purple splotches, stipules deeply divided (0.6–1.0 cm long)

flowers   in small rounded clusters, 2- to 5-flowered; calyx tube pubescent (1 mm long); corolla yellow (4–5 mm long)

fruit   legume, 2–5 times spirally coiled, with spines (2–3 mm long) in a double row

other   stems puberulent to essentially glabrate

HISTORIC, FOOD, AND MEDICINAL USES   young leaves may be used to garnish salads

LIVESTOCK LOSSES   excessive grazing of fresh foliage may cause bloating, wool becomes infested with burs

FORAGE VALUE   valuable annual forage for livestock, burs are consumed during the dry season

HABITAT   rich valley loams or lower slopes of foothills, waste grounds, common lawnweed

Plains loco (*Oxytropis campestris*)

Family	FABACEAE
Species	*Oxytropis campestris* (L.) DC.
Common Name	Plains loco
Life Span	Perennial
Origin	Native
Season	Cool

GROWTH CHARACTERISTICS   acaulescent forb (scapes 20–50 cm tall), grows May–July

DISTINGUISHING CHARACTERISTICS

leaves   pinnately compound, leaflets (up to 31) oblong-lanceolate (1.0–2.5 cm long), appressed simple hairs, stipules adnate to petioles

flowers   many-flowered capitate or oblong racemes; calyx (5–7 mm long), appressed villous, hairs often blackish; corolla (1.2–1.5 cm long) cream-colored to yellowish, keel petals may have purple blotches

fruit   legume, oblong-ovate (1.5–2.0 cm long), erect to spreading, suture strongly intruded, nearly 2-celled

other   pubescence of herbage extremely variable, hairs simple

HISTORIC, FOOD, AND MEDICINAL USES   none

LIVESTOCK LOSSES   may cause loco in horses, cattle, sheep, and goats on range and therefore care should be taken when grazing infested areas

FORAGE VALUE   poor to worthless, although generally not consumed

HABITAT   plains, woods, and meadows to rocky and gravelly slopes

343

Lambert crazyweed (*Oxytropis lambertii*)

Family	FABACEAE
Species	*Oxytropis lambertii* Pursh
Common Name	Lambert crazyweed (white loco, white point loco)
Life Span	Perennial
Origin	Native
Season	Cool

GROWTH CHARACTERISTICS   forb (scapes 10–35 cm tall), acaulescent, tufted scapes, grows from April–July, reproduces by seeds

DISTINGUISHING CHARACTERISTICS

leaves   pinnately compound, 7–17 leaflets, narrow to oblong (1.0–3.5 cm long), pubescent, stipules adnate to petioles

flowers   loose-flowered racemes; calyx tube silky-pilose (6–8 mm long); corolla pink-purple to various shades of rose (2.2 cm long), keel petal with an abrupt point

fruit   legume, ovoid to oblong-ovoid (0.7–1.0 cm long), with a beak (3–5 mm long)

other   herbage pubescent throughout, hairs attached at middle, with 2 free ends

HISTORIC, FOOD, AND MEDICINAL USES   famous in the history of the West as one of the causes of "locoed" horses

LIVESTOCK LOSSES   poisonous, can cause loco disease in horses, cattle, sheep, and goats, caused by the accumulation of alkaloids; the common name crazyweed is derived from the erratic behavior of animals feeding on this plant, plant readily eaten by livestock

FORAGE VALUE   poor to worthless, animals may utilize Lambert crazyweed when other forage is not available

HABITAT   dry upland prairie and plains

Purple prairie clover (*Petalostemum purpureum*)

Family	FABACEAE
Species	*Petalostemum purpureum* (Vent.) Rydb.
Common Name	Purple prairie clover
Life Span	Perennial
Origin	Native
Season	Warm

GROWTH CHARACTERISTICS   forb (to 60 cm tall), branched or simple, erect or ascending, flowers in June–July, reproduces by seeds and root-stocks

DISTINGUISHING CHARACTERISTICS

leaves   alternate, pinnately compound, leaflets linear (mostly 5), (0.8–2.0 cm long), usually folded, glabrous to villous, glandular-punctate below

flowers   irregular, in dense spikes (1–5 cm long); calyx tube (2.5–4.0 mm long), lobes shorter than the tube, silky-villous; corolla rose-purple

fruit   legume (3 mm long), pubescent

other   woody taproot

HISTORIC, FOOD, AND MEDICINAL USES   American Indians ate fresh leaves and boiled leaves

LIVESTOCK LOSSES   may cause bloat

FORAGE VALUE   excellent for livestock and wildlife, important component of prairie hay, high in protein, highly palatable and nutritious

HABITAT   plains and hills

Honey mesquite (*Prosopis glandulosa*)

Notes:

Family	FABACEAE
Species	*Prosopis glandulosa* Torr.
Common Name	Honey mesquite (mesquite, glandular mesquite, mezquite)
Life Span	Perennial
Origin	Native
Season	Warm

GROWTH CHARACTERISTICS   shrub (1–3 m tall), deciduous, much-branched; growth begins in late spring, flowers May, fruit matures August; growth and reproduction is regulated by moisture availability when temperatures are warm

DISTINGUISHING CHARACTERISTICS

leaves   bipinnately compound, alternate, 6–30 pairs of leaflets; blades linear to linear-oblong (3–5 cm long), glabrous or nearly so; stipules modified as spines

flowers   yellow in spike-like inflorescence, petals are pubescent within, pedicel glandular

fruit   legume (10–20 cm long), linear, straight or nearly so, slight constriction between seeds, pedicel glandular

other   armed with stipular spines (to 5 cm long), rigid and straight, or sometimes spineless; twigs have zig-zag appearance

HISTORIC, FOOD, AND MEDICINAL USES   wood is used for fuel and lumber; beans were important in diets of Southwestern Indians, they also prepared flour from legumes, fermentation of this flour produced intoxicating beverage

LIVESTOCK LOSSES   ingestion of large amounts may result in rumen stasis

FORAGE VALUE   foliage and legumes provide poor-good forage for livestock, deer and javelina, seeds important for food source for numerous wildlife species

HABITAT   plains and prairies; especially abundant on disturbed grasslands

Slimflower scurfpea (*Psoralea tenuiflora*)

Notes:

Family	FABACEAE
Species	*Psoralea tenuiflora* Pursh
Common Name	Slimflower scurfpea (wild alfalfa, slender scurfpea)
Life Span	Perennial
Origin	Native
Season	Warm

GROWTH CHARACTERISTICS   forb (20−60 cm tall), much-branched, reproduces from seeds and long creeping rootstalks, drought resistant

DISTINGUISHING CHARACTERISTICS

leaves   alternate, palmately compound, usually 3 (sometimes 5) leaflets, leaflets linear-oblanceolate (1−4 cm long), pubescent above, glabrate below, glandular dotted on both surfaces

flowers   solitary or 3 at a node; calyx tube glandular (2.0−2.5 mm long), corolla blue to purple (4−7 mm long)

fruit   legume (5−8 mm long), ovoid, with a short straight beak, glandular-dotted

other   the above ground portion disarticulates at the crown following maturity and tumbles with the wind

HISTORIC, FOOD, AND MEDICINAL USES   American Indians drank a tea from stems and leaves for fever

LIVESTOCK LOSSES   reported to be poisonous to cattle and horses, but no experimental or circumstantial evidence is available in support of this report; may cause bloat

FORAGE VALUE   low palatability and value when green, after curing it is readily eaten

HABITAT   dry plains, prairies, and open woods

Gambel oak (*Quercus gambelii*)

Family	FAGACEAE
Species	*Quercus gambelii* Nutt.
Common Name	Gambel oak (encinco)
Life Span	Perennial
Origin	Native
Season	Cool

GROWTH CHARACTERISTICS   shrub or small tree (to 20 m tall), deciduous, grows in dense stands, often forming thickets

DISTINGUISHING CHARACTERISTICS

leaves   alternate, simple (8–16 cm long and 4.5–7.0 cm wide), usually broadest above the middle, deeply 5–9 lobed, lobes rounded, tapering above to squared at the base, lobed ½ way to midrib, leaves are leathery yellow-green above, pale and smooth or densely hairy beneath

flowers   unisexual, plant monoecious; in catkins

fruit   acorn, solitary or clustered, broadest at base to nearly globe-shaped, cup shallowly to deeply bowl-shaped, enclosing ¼ – ⅓ of the nut, scales hairy

other   bark thin, light gray to white, becoming scaly with age

HISTORIC, FOOD, AND MEDICINAL USES   acorns may be eaten after tannic acid is removed, American Indians used acorns to thicken soup and make mush, they also obtained oil from acorns to be used as flavoring

LIVESTOCK LOSSES   young shoots contain 4–10% tannic acid, therefore poisoning of cattle and occasionally sheep is common from March–April

FORAGE VALUE   leaves are fair for all classes of livestock, deer, and porcupines; acorns eaten by livestock and wildlife

HABITAT   dry foothills, canyons, and lower mountain slopes

Blackjack oak (*Quercus marilandica*)

Notes:

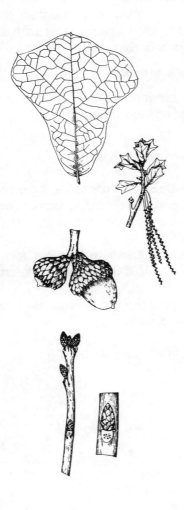

354

Family	FAGACEAE
Species	*Quercus marilandica* Muenchh.
Common Name	Blackjack oak (encinco)
Life Span	Perennial
Origin	Native
Season	Warm

GROWTH CHARACTERISTICS   shrub or tree (to 12 m tall), deciduous, flowers and leaves appear in May, reproduces by seeds and sprouts

DISTINGUISHING CHARACTERISTICS

leaves   alternate, leathery, margin revolute, broadly obovate to bat-shaped (10–25 cm long and almost as wide), apex 3-lobed or sometimes merely dentate, midveins of lobes end as a bristle, upper surface dark green and glossy, lower surface glabrous to yellow-hairy along midrib

flowers   unisexual, in staminate and pistillate catkins

fruit   acorns enclosed in cups for ½–⅔ their length, cup light yellow to brown with red-brown scales, nut oblong, acorns ripen in 2 years

other   bark dark brown or black, deeply fissured

HISTORIC, FOOD, AND MEDICINAL USES   wood used for posts, fuel, charcoal, and railroad ties

LIVESTOCK LOSSES   tannic acid poisoning can be a problem for cattle

FORAGE VALUE   buds, twigs, and leaves browsed by cattle and goats; acorns provide food, and tree provides cover for squirrels, turkeys, and deer

HABITAT   dry, sandy, and sterile soils

Post oak (*Quercus stellata*)

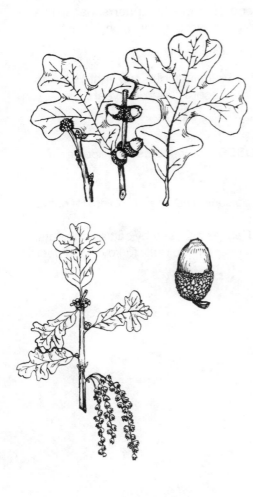

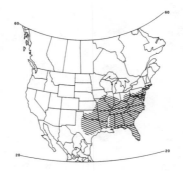

Family	FAGACEAE
Species	*Quercus stellata* Wang.
Common Name	Post oak (encinco)
Life Span	Perennial
Origin	Native
Season	Warm

GROWTH CHARACTERISTICS   shrub or tree (10–30 m tall), deciduous, flowers March–May, fruit matures September–November, reproduces by seeds or sprouts

DISTINGUISHING CHARACTERISTICS

leaves   alternate (10–15 cm long and 7.5–10.0 cm wide), leathery, broadest above the middle with 3–5 short deep wide lobes making it somewhat cross-shaped, lower lobes deep (more than half way to midvein), base is tapered and apex rounded, gray-pubescence in axils of leaf veins on lower surface, stellate hairs on lower surface

flowers   unisexual, in staminate pistillate catkins

fruit   acorns in clusters of 2–4, cup is top-shaped covering about ⅓ of the nut

other   bark dark gray, rough with scaly ridges

HISTORIC, FOOD, AND MEDICINAL USES   wood used for furniture, fencing, flooring, fuel, railroad ties, and lumber

LIVESTOCK LOSSES   tannic acid poisoning may be a problem for cattle

FORAGE VALUE   buds, leaves, and twigs eaten by cattle and goats in early spring; acorns eaten by deer, wild turkeys, raccoons, and squirrels

HABITAT   rocky and sandy ridges

Broadleaf filaree (*Erodium botrys*)

Notes:

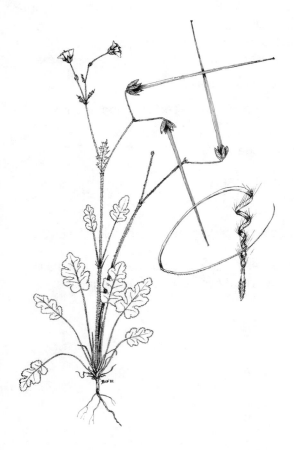

Family	GERANIACEAE
Species	*Erodium botrys* (Cav.) Bert.
Common Name	Broadleaf filaree
Life Span	Annual
Origin	Introduced (from Mediterranean region through Europe)
Season	Cool

GROWTH CHARACTERISTICS   forb (stems 10–90 cm long), prostrate to somewhat erect, first forming a rosette, germinates in winter and grows rapidly for about 3–4 months

DISTINGUISHING CHARACTERISTICS

leaves   opposite, ovate to oblong ovate (3–8 cm long), shallowly lobed or pinnatifid, bristly-hairy on veins and margins, stipulate

flowers   perfect, 5-merous; sepals (7–8 mm long, enlarging in fruit) with red awn-tips; petals lavender (1.5 cm long); style-column coiled with numerous turns (9.5–12.5 cm long)

fruit   indehiscent carpel bodies (0.8–1.0 cm long), pubescent

other   plant with recurved stiff pubescence

HISTORIC, FOOD, AND MEDICINAL USES   none

LIVESTOCK LOSSES   may cause bloat

FORAGE VALUE   good early forage, especially valuable on annual ranges

HABITAT   grassy lowlands and foothills

Red stem filaree (*Erodium cicutarium*)

Notes:

Family	GERANIACEAE
Species	*Erodium cicutarium* (L.) L'Her.
Common Name	Redstem filaree (storksbill, heronbill, alfilaria alfilerillo)
Life Span	Annual
Origin	Introduced (from Mediterranean region through Europe)
Season	Cool

GROWTH CHARACTERISTICS   forb (stems 10–50 cm long), decumbent, acaulescent at first then elongating, one of the first plants to germinate in late fall or spring

DISTINGUISHING CHARACTERISTICS

leaves   opposite (3–10 cm long), margins deeply cut, lobes acute, often toothed, with pedicels (0.8–1.8 cm long), glandular-pubescent, stipulate

flowers   perfect, 5-merous; sepals (3–5 mm long) with short white awn-points; petals rose-lavender (5–7 mm long) ciliate at the base; style column (2–4 cm long) coiled with several turns

fruit   5 indehiscent carpel bodies (4–5 mm long); stiffly pubescent

HISTORIC, FOOD, AND MEDICINAL USES   young leaves can be eaten raw or cooked; contains an antidote for strychnine

LIVESTOCK LOSSES   may cause bloat

FORAGE VALUE   excellent to good spring forage for cattle, sheep, and wildlife

HABITAT   open, cultivated dry places, waste grounds, plains, and mesas

Wild white geranium (*Geranium richardsonii*)

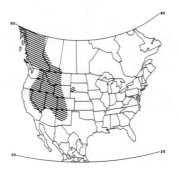

Family	GERANIACEAE
Species	*Geranium richardsonii* Fisch. & Trautv.
Common Name	Wild white geranium (Richardson geranium)
Life Span	Perennial
Origin	Native
Season	Warm

GROWTH CHARACTERISTICS   forb (0.5–1.0 m tall), from a thick root-stock, stems 1–few, flowers June–July, fruits mature August–September, reproduces from seeds

DISTINGUISHING CHARACTERISTICS

leaves   basal leaves on long petioles, blades palmately 5- to 7-lobed (5–30 cm long and 3–15 cm wide), upper leaves opposite, reduced in size, on petioles (2–12 cm long), upper pedicels possessing hairs tipped with purple glands

flowers   regular, 5-merous, fused at base; petals (1.0–1.8 cm long), white to pink with purple veins; calyx (5–9 mm long), awn-tipped

fruit   carpel bodies (3–4 mm long) with 3–5 style branches, somewhat pubescent, seeds (2.5–3.5 mm long)

other   stems pubescent

HISTORIC, FOOD, AND MEDICINAL USES   Cheyenne Indians used pulverized leaf blades to treat nosebleeds, a powder was made and snuffed, the roots were also powdered and made into a drink

LIVESTOCK LOSSES   none

FORAGE VALUE   worthless to poor for cattle, poor to fair for sheep, flowers and leaves grazed by deer

HABITAT   mountains and foothills

St. Johnswort (*Hypericum perforatum*)

Family	HYPERICACEAE
Species	*Hypericum perforatum* L.
Common Name	St. Johnswort (Klamathweed, goatweed)
Life Span	Perennial
Origin	Introduced (from Europe and Africa)
Season	Warm

GROWTH CHARACTERISTICS   forb (0.3–1.5 m tall), leafy basal off-shoots, stems much-branched

DISTINGUISHING CHARACTERISTICS

leaves   opposite on main axis, linear oblong (2–4 cm long), diverge from stem at a 90° angle, subtended by short leafy branchlets, margins en-rolled, surfaces covered with small black glands

flowers   cymes, densely flowered; 5 sepals, linear-lanceolate, persistent; 5 petals orange-yellow, often black-dotted (0.8–1.2 cm long), twisting when dry; stamens numerous

fruit   capsules containing several small, brown seeds

other   stems are red and appear "jointed" due to opposite leaf scars

HISTORIC, FOOD, AND MEDICINAL USES   Menominee Indians mixed St. Johnswort with black raspberry root in hot water and drank the tea for tuberculosis, also acts as a diuretic, and may kill internal worms

LIVESTOCK LOSSES   poisonous to livestock, causes a photosensitiz-ing reaction resulting in dermatitis to nonpigmented skin (especially in light-colored horses, cattle, and sheep); sunlight acts as a catalyst, therefore symptoms only occur when livestock are exposed to strong sunlight after grazing

FORAGE VALUE   poor to worthless for livestock and wildlife, except it is fair for goats

HABITAT   waste areas, disturbed fields, pastures, and roadsides

365

Range ratany (*Krameria parvifolia*)

Family	KRAMERIACEAE
Species	*Krameria parvifolia* Benth.
Common Name	Range ratany (little ratany, littleleaf krameria)
Life Span	Perennial
Origin	Native
Season	Cool

GROWTH CHARACTERISTICS
shrub (to 60 cm tall), deciduous, low, much-branched, flowers April–May, reproduces by seeds

DISTINGUISHING CHARACTERISTICS

leaves   alternate, linear (0.3–1.5 cm long), silky-pubescent, appearing light gray in color, stipulate

flowers   axillary on hairy-glandular peduncles; corolla purple and showy, composed of 5 petals with 3 united to form a short claw; stamens 3–4, partially united

fruit   rounded pod, silky-hairy with slender barbed spines, bur-like

other   stems densely gray-hairy when young, becoming brown to black with age

HISTORIC, FOOD, AND MEDICINAL USES
American Indians made a decoction to use as an eyewash, for diarrhea, and for sores; source for red or brown dye; roots used in manufacturing ink

LIVESTOCK LOSSES
none

FORAGE VALUE
valuable browse for all livestock classes, cannot withstand heavy grazing because branches are brittle, fruit is bur-like and is readily disseminated by livestock, important browse for mule deer

HABITAT
dry rocky slopes and plains, sandy hillsides

Scarlet globemallow (*Sphaeralcea coccinea*)

Notes:

Family	MALVACEAE
Species	*Sphaeralcea coccinea* (Pursh) Rydb.
Common Name	Scarlet globemallow (red falsemallow)
Life Span	Perennial
Origin	Native
Season	Warm

GROWTH CHARACTERISTICS   forb (10–30 cm tall), from a stout tap-root, erect to decumbent, flowers in early spring, reproduces by seeds

DISTINGUISHING CHARACTERISTICS

leaves   alternate, petiolate, deeply cleft, deltoid in outline (1–6 cm long and wider than long), 3-several palmate lobes, covered with stellate pubescence

flowers   calyx of 5 sepals, conspicuously villous (0.5–1.0 cm long); 5 petals, orange or scarlet to salmon-colored (1–2 cm long); stamens numerous

fruit   carpel differentiated into a smooth dehiscent portion at apex and roughened indehiscent basal portion, 1 seed per carpel

other   entire plant is covered with stellate pubescence

HISTORIC, FOOD, AND MEDICINAL USES   Blackfoot Indians chewed the plant and applied the paste to burns, scalds, and sores as a cooling agent

LIVESTOCK LOSSES   none

FORAGE VALUE   excellent for deer, fair to little palatability for domestic livestock; important forage in Southwest, not grazed much in Central Plains; increases in abundance in overgrazed areas and during dry periods

HABITAT   dry hills and plains, roadsides

Fireweed (*Epilobium angustifolium*)

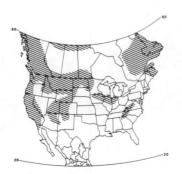

Family	ONAGRACEAE
Species	*Epilobium angustifolium* L.
Common Name	Fireweed (willowherb, blooming Sally)
Life Span	Perennial
Origin	Native
Season	Warm

GROWTH CHARACTERISTICS   forb (0.6–2.5 m tall), reproduces by seeds and underground rootstocks, generally not branched, flowers July–August, especially abundant after forest fires (hence common name)

DISTINGUISHING CHARACTERISTICS

leaves   alternate, sessile, lanceolate (5–20 cm long), pale beneath, veins are united near leaf margin, evident midrib, margin entire, apex acuminate

flowers   arranged in terminal racemes; 4 sepals (0.8–1.2 cm long) lanceolate-linear, often tinged in purple, canescent; 4 petals (0.8–1.8 cm long), lilac-purple to rose-colored, clawed; 8 stamens shorter than petals; stigma 4-lobed, these long and slender, ovary inferior

fruit   capsule (5–8 cm long), canescent, often purplish; seeds oblong (1.0–1.4 mm), numerous, with a tuft of silky hairs

HISTORIC, FOOD, AND MEDICINAL USES   used as a potherb, young shoots can be cooked like asparagus, young leaves used in salads, steeped for tea, pith of stem can be used to flavor and thicken stews and soups; grown as an ornamental

LIVESTOCK LOSSES   none

FORAGE VALUE   good for sheep, fair for cattle; grazed to some extent by horses, deer, and elk

HABITAT   disturbed areas, moist places, open woods

Pinyon pine (*Pinus edulis*)

Family	PINACEAE
Species	*Pinus edulis* Engelm.
Common Name	Pinyon pine (Mexican piñon pine, pino pinonero)
Life Span	Perennial
Origin	Native
Season	Evergreen

GROWTH CHARACTERISTICS   tree, much-branched (5–14 m tall), reproduces by seeds

DISTINGUISHING CHARACTERISTICS

leaves   needles (1.2–5.0 cm long), 2 per fascicle, stiff, sharp-pointed bluish-green when growth is fresh, turning yellowish-green

flowers   male cones dark red, small near tips, at end of branches in clusters; female cones purple, also at ends of branches

fruit   egg-shaped cone (3.8–5.0 cm long); scales thick, short, and blunt; seeds egg-shaped (1.2–1.4 cm long)

other   bark thin, gray to reddish-brown

HISTORIC, FOOD, AND MEDICINAL USES   the seed crop is valuable and is used in making confectionery; seeds were staple food in American Indian diet, eaten raw, roasted, or ground into flour; needles were steeped for tea; inner bark served as starvation food for American Indians; wood is used for fuel and fence posts

LIVESTOCK LOSSES   none

FORAGE VALUE   seeds important wildlife food for several song birds, quail, squirrels, chipmunks, black bears, mule deer, and goats

HABITAT   mesas, plateaus, and foothills

Ponderosa pine (*Pinus ponderosa*)

Notes:

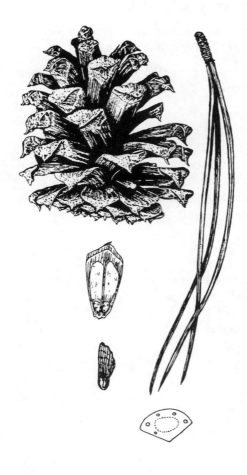

Family	PINACEAE
Species	*Pinus ponderosa* Laws.
Common Name	Ponderosa pine (western yellow pine, pino ponderosa)
Life Span	Perennial
Origin	Native
Season	Evergreen

GROWTH CHARACTERISTICS   tree (30–60 m tall), open and often round-topped crown, trunk tall and straight, older trees generally do not have lower branches

DISTINGUISHING CHARACTERISTICS

leaves   needles, 2 or 3 per fascicle, generally 3 (10–28 cm long), form brush-like tufts at ends of naked branches, bright yellowish green to dark green, fascicle sheath dark brown to black, becoming rough with age

flowers   produced in cones, staminate cones in clusters at base of new growth, cylinder-shaped female cones clustered or in pairs nearly globe-shaped

fruit   woody cones, broadest at base or near middle (7.6–12.5 cm long), scales armed with a spine, light reddish-brown; seeds dark-brown to pur-plish-mottled (7–8 mm long)

other   bark is rough, reddish-brown to cinnamon colored scales darkening with age; buds chestnut brown, resinous

HISTORIC, FOOD, AND MEDICINAL USES   American Indians used cones to make quick fires, wood used commercially for boxes, crates, construction, and mill products; used to some extent in shelterbelts

LIVESTOCK LOSSES   browsing on needles may cause abortion

FORAGE VALUE   seeds are eaten by several bird species and small mammals; browsed by mule deer, white-tailed deer, and mountain sheep

HABITAT   mountain ranges and plateaus, in a wide variety of soil types

Woolly indianwheat (*Plantago patagonica*)

Family	PLANTAGINACEAE
Species	*Plantago patagonica* Jacq.
Common Name	Woolly indianwheat (woolly plantain)
Life Span	Annual
Origin	Native
Season	Cool

GROWTH CHARACTERISTICS   forb (10–20 cm tall), acaulescent, starts growth in late winter with favorable rainfall, abundant following wet winters, reproduces by seeds

DISTINGUISHING CHARACTERISTICS

leaves   basal, linear (3–10 cm long and 1–4 mm wide), margins entire, covered with woolly pubescence

flowers   in dense spikes (1–2 cm long), flowers small and perfect; corolla lobes white (1–2 mm long), spreading

fruit   capsule with 2 seeds, breaking apart at middle

other   whole plant appears silvery white due to pubescence, fine taproot

HISTORIC, FOOD, AND MEDICINAL USES   American Indians chewed and swallowed leaves for internal hemorrhage, leaves were also chewed for toothache

LIVESTOCK LOSSES   none

FORAGE VALUE   good for sheep and fair for cattle, may be a major forage species on lambing ranges, abundance is generally an indicator of deteriorated conditions, in Southwest it is often regarded as a spring opportunist

HABITAT   plains and slopes, generally in sandy soils

Low larkspur (*Delphinium bicolor*)

Family	RANUNCULACEAE
Species	*Delphinium bicolor* Nutt.
Common Name	Low larkspur (little larkspur)
Life Span	Perennial
Origin	Native
Season	Cool

GROWTH CHARACTERISTICS   forb (15–30 cm tall), from a branching cluster of fascicled roots, growth begins as snow melts in early spring, matures by June–July, reproduces by seeds

DISTINGUISHING CHARACTERISTICS

leaves   mostly basal, wedge-shaped in outline (2–4 cm wide), dissected into linear-oblong divisions, puberulent to glabrate

flowers   in racemes; sepals petaloid (1.5 cm long), upper ones brown to yellow with blue veins, lower ones violet with wavy lobes

fruit   follicle, densely viscid-pubescent to glabrous (1.5–2.0 cm long)

other   stems finely hairy to glabrous

HISTORIC, FOOD, AND MEDICINAL USES   American Indians crushed plants of this genus and applied it to their hair to control lice and other insects

LIVESTOCK LOSSES   poisonous to cattle throughout growth cycle, contains alkaloids

FORAGE VALUE   fair to good for sheep and some wildlife

HABITAT   open hills and meadows, dry soils

Tall larkspur (*Delphinium occidentale*)

Notes:

Family	RANUNCULACEAE
Species	*Delphinium occidentale* (Wats.) Wats.
Common Name	Tall larkspur (duncecap larkspur)
Life Span	Perennial
Origin	Native
Season	Warm

GROWTH CHARACTERISTICS   forb (0.6–2.0 m tall), erect, from a woody root stalk, growth starts in late spring, flowers July–August, seed matures August–September, seems to be concentrated where snow pack lasts the longest

DISTINGUISHING CHARACTERISTICS

leaves   alternate (10–18 cm long), on petioles, blades wedge-shaped (8–12 cm wide), palmately divided, 3–5 divisions

flowers   arranged in racemes (15 cm long), rachis glandular-hairy; flowers irregular; sepals 5, petaloid (0.6–1.2 cm long), dark bluish-purple; spur horizontal (0.9–1.2 cm long)

fruit   many-seeded follicle (0.9–1.2 cm long), glabrous to glandular-pubescent

other   stem hollow, straw-colored, especially at base

HISTORIC, FOOD, AND MEDICINAL USES   American Indians crushed plants of this genus and applied it to their hair to control lice and other insects

LIVESTOCK LOSSES   poisonous to cattle until after blossoming, contains alkaloids

FORAGE VALUE   fair to good for sheep and some wildlife

HABITAT   dry slopes, meadows, thickets, and streambanks

381

Wedgeleaf ceanothus (*Ceanothus cuneatus*)

382

Family	RHAMNACEAE
Species	*Ceanothus cuneatus* (Hook.) Nutt.
Common Name	Wedgeleaf ceanothus (buckbrush, narrowleaf buckbrush)
Life Span	Perennial
Origin	Native
Season	Warm

GROWTH CHARACTERISTICS   rigid shrub (1.0–3.5 m tall), may form dense thickets, flowers March–May, fire and drought resistant

DISTINGUISHING CHARACTERISTICS

leaves   simple (0.5–2.0 cm long), persistant, opposite or whorled, narrowly oblong-cuneiform, finely tomentose canescent beneath, green and glabrous above, leather-like, apex obtuse, margins entire to finely toothed at apex

flowers   white, small but showy, in axillary umbellate clusters, corolla of 5 petals; calyx of 5 sepals united at base

fruit   capsule, subglobose (5–6 mm broad), short erect horns near top; seeds black, shiny

other   branches thorny and rigid, with grayish pubescence

HISTORIC, FOOD, AND MEDICINAL USES   leaves and flowers may be boiled for tea; infusion from bark used to make tonic; fresh flowers when crushed and rubbed in water make a perfumed, cleansing lather

LIVESTOCK LOSSES   none

FORAGE VALUE   browsed by cattle when other forage is unavailable, browse for sheep, goats and deer; seeds eaten by squirrels

HABITAT   dry gravelly ridges and slopes, open rocky sites

Fendler ceanothus (*Ceanothus fendleri*)

Family	RHAMNACEAE
Species	*Ceanothus fendleri* Gray
Common Name	Fendler ceanothus (deerbriar, buckbrush, Fendler soapbloom)
Life Span	Perennial
Origin	Native
Season	Warm

GROWTH CHARACTERISTICS   shrub (1 m or less tall), deciduous, loosely branched, flowers April–October

DISTINGUISHING CHARACTERISTICS

leaves   simple (1.0–2.5 cm long), alternate, short petiolate, close glandular, serrate to entire, distinctly 3-veined, elliptic, white silky-hairy beneath, greener above

flowers   small, white and numerous in showy clusters at or near the ends of twigs; corolla of 5 petals clawed and hooded; calyx united below

fruit   capsule, globular in shape with 3 lobes near the top

other   branches often blue-gray or glaucous, branches may be sharp-pointed to thorn-tipped, twigs are green when young

HISTORIC, FOOD, AND MEDICINAL USES   American Indians extracted a dye from this plant

LIVESTOCK LOSSES   none

FORAGE VALUE   important browse for goats, fair to good for cattle and sheep, frequently browsed by horses; heavily utilized by porcupines in summer months

HABITAT   foothills and mountains on dry, well-drained soils

Deerbrush (*Ceanothus integerrimus*)

Notes:

Family	RHAMNACEAE
Species	*Ceanothus integerrimus* H. & A.
Common Name	Deerbrush (bluebrush, mountain lilac, sweet birch)
Life Span	Perennial
Origin	Native
Season	Warm

GROWTH CHARACTERISTICS   shrub (1–4 m tall), deciduous, loosely branched, branches often drooping, flowers May–July, reproduces by seeds and rootstalks, increases following fire

DISTINGUISHING CHARACTERISTICS

leaves   simple (2.5–7.0 cm long), alternate, petioled, entire to denticulate near tip, elliptical to oval, rounded at base, light green above, paler and mostly pubescent beneath, 3-veined from base

flowers   white to dark blue, sweet-scented in terminal clusters (4–15 cm long); corolla of 5 clawed and hooded petals; calyx whitish of 5 united sepals; stamens 5, inserted opposite from petals

fruit   pear-shaped capsules, viscid, 3-celled with 3 lobes

other   branches often green to yellow, sometimes warty below; flowers in bud with scaly bracts

HISTORIC, FOOD, AND MEDICINAL USES   valuable honey plant

LIVESTOCK LOSSES   none

FORAGE VALUE   valuable browse; good to excellent for cattle, sheep, goats, and deer; fair to good for horses

HABITAT   well-drained, fertile soils on dry slopes and ridges

Snowbrush (*Ceanothus velutinus*)

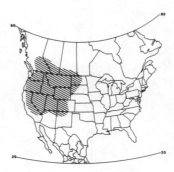

388

Family	RHAMNACEAE
Species	*Ceanothus velutinus* Dougl. *ex* Hook.
Common Name	Snowbrush (tobaccobrush, mountain balm)
Life Span	Perennial
Origin	Native
Season	Evergreen

GROWTH CHARACTERISTICS   shrub (1–2 m tall), diffuse-spreading, often round-topped, flowers May–July, germination is stimulated by burning

DISTINGUISHING CHARACTERISTICS

   leaves   simple, alternate, petiolate, broadly ovate to elliptical, (2.5–8.0 cm long), dark green, glabrous to viscid above (sometimes appearing "varnished"), margin closely glandular-serrulate

   flowers   white, in clusters (5–10 cm long); corolla of 5 petals

   fruit   capsule, subglobose to triangular (3–4 mm long), 3-lobed at summit

   other   strong balsamic odor, no spines

HISTORIC, FOOD, AND MEDICINAL USES   some American Indians used leaves as a tobacco substitute

LIVESTOCK LOSSES   none

FORAGE VALUE   generally not considered a browse species, although it is occasionally browsed by deer and elk in winter; goats, sheep, and deer may graze blossoms; deer often bed in this shrub; nitrogen-fixing by root nodules

HABITAT   open wooded slopes; pioneer, forming vast thickets in logged or burned areas

Chamise (*Adenostoma fasciculatum*)

Family	ROSACEAE
Species	*Adenostoma fasciculatum* H. & A.
Common Name	Chamise (greasewood, yerba del pasmo)
Life Span	Perennial
Origin	Native
Season	Evergreen

GROWTH CHARACTERISTICS   shrub (0.5–3.5 m tall), forms dense thickets, burl sprouter, growth starts in January and ends in June

DISTINGUISHING CHARACTERISTICS

leaves   fascicled, entire, awl-shaped to club-shaped, channeled on one side, petiolate (0.4–1.0 cm long) small stipules present, glabrous, often resinous; seedling leaves are lobed or divided, usually single on young stems

flowers   floral tube small, green; in panicles (4–12 cm long)

fruit   achene, small, hard, 1-seeded

other   bark is red, becoming shreddy with age; diffusely branched with some slender stems tapering to spine-like points

HISTORIC, FOOD, AND MEDICINAL USES   used by some Indian tribes for various ceremonial purposes

LIVESTOCK LOSSES   none

FORAGE VALUE   largely unpalatable for livestock, good browse for deer; provides watershed protection

HABITAT   dry slopes, ridges, and foothills; chaparral

Serviceberry (*Amelanchier alnifolia*)

Notes:

392

Family	ROSACEAE
Species	*Amelanchier alnifolia* (Nutt.) Nutt.
Common Name	Serviceberry (Saskatoonberry, juneberry)
Life Span	Perennial
Origin	Native
Season	Cool

GROWTH CHARACTERISTICS   shrub or small tree (to 6 m tall), deciduous, flowers from May–June, fruit in July–August, forming colonies

DISTINGUISHING CHARACTERISTICS

leaves   simple, alternate, elliptic to oval with square cut tips (1–4 cm long and as wide as long), petiolate, serrate or dentate above middle, dark green above, finely tomentose beneath, not leathery, no prominent veins

flowers   white, in small racemes, 3- to 6-flowered, ill-scented

fruit   berries yellow, reddish, or purplish-black, rounded to egg-shaped (7–8 mm long), fleshy

other   young growth white-pubescent; rigid twigs; bark dark gray to brown

HISTORIC, FOOD, AND MEDICINAL USES   fruits can be used to make jams, pies, and wines; American Indians used stems for arrow shafts and tepee stakes

LIVESTOCK LOSSES   none

FORAGE VALUE   young growth is fair to good browse for livestock, excellent browse for deer and moose; fruit is an important food for chipmunks, squirrels, black bears, and birds; bark is eaten by beaver and marmots

HABITAT   dry and rocky slopes to moist and fertile soils

Curlleaf mountain mahogany (*Cercocarpus ledifolius*)

Notes:

Family	ROSACEAE
Species	*Cercocarpus ledifolius* Nutt. *ex* Torr. & Gray
Common Name	Curlleaf mountain mahogany (desert mahogany)
Life Span	Perennial
Origin	Native
Season	Evergreen

GROWTH CHARACTERISTICS   shrub or small tree (1−5 m tall), flowers May−July, reproduces by seeds

DISTINGUISHING CHARACTERISTICS

leaves   alternate, petiolate, narrowly lance-shaped, (1.5−3.0 cm long and 0.8−1.7 cm wide), apex sharp-pointed, base wedge-shaped, leathery, margin entire, involute, dark green above, paler and with rusty hair beneath, resinous, 1 prominent midrib

flowers   single in leaf axils; calyx united to form a tube (4−9 cm long); no petals; 20−30 stamens; single pistil, style elongating in fruit

fruit   achene, hard, narrowed, sharp-pointed (5−7 mm long), tipped with a persistent feathery style (5−8 cm long)

other   bark reddish-brown becoming deeply furrowed

HISTORIC, FOOD, AND MEDICINAL USES   wood is hard and dense (will not float) providing excellent fuel, giving off intense heat and burns for long periods; Gosiute Indians of Utah made bows from wood

LIVESTOCK LOSSES   none

FORAGE VALUE   fair for sheep and cattle in fall and winter, good for big game in winter, provides cover and some browse for deer

HABITAT   gravelly slopes and rocky ridges at high altitudes

True mountain mahogany (*Cercocarpus montanus*)

Notes:

Family	ROSACEAE
Species	*Cercocarpus montanus* Raf.
Common Name	True mountain mahogany
Life Span	Perennial
Origin	Native
Season	Cool

GROWTH CHARACTERISTICS　　　shrub or small tree (1–6 m tall), strict or spreading, reproduces by seeds, fruit matures by August

DISTINGUISHING CHARACTERISTICS

leaves　　simple (2–5 cm long and 1.5–3.5 cm wide), alternate or fascicled, petiolate, thin to somewhat thick and firm, lanceolate to oblanceolate, broadest above the middle, wedge-shaped at base and rounded at apex, green to gray-green above, densely hairy (lighter) beneath, lateral veins 3–10, evident beneath, stipulate, margin coarsely serrate

flowers　　tubular, solitary, or fascicled in leaf axils

fruit　　slender villous achene, style long (6–10 cm), plumose and persistent, twisted and much exserted

other　　bark gray to brown, thin

HISTORIC, FOOD, AND MEDICINAL USES　　　American Indians used this plant to make tools and war clubs

LIVESTOCK LOSSES　　　may contain cyanogenetic glycoside

FORAGE VALUE　　　good to very good browse for all types of livestock; extremely valuable winter browse for deer

HABITAT　　　dry, rocky bluffs and mountainsides

Blackbrush (*Coleogyne ramosissima*)

Family	ROSACEAE
Species	*Coleogyne ramosissima* Torr.
Common Name	Blackbrush (burrobrush)
Life Span	Perennial
Origin	Native
Season	Evergreeen

GROWTH CHARACTERISTICS   shrub (0.5–5.0 m tall), much branched; the branches opposite, short, rigid, frequently ending in a spinelike twig; flowers April–May when moisture is available

DISTINGUISHING CHARACTERISTICS

leaves   opposite, in fascicles, (4–11 mm long) awl-shaped, entire, coriaceous, apex obtuse to mucronate; 4-grooved beneath, canescent, 2-branched hairs appressed to the surface of the leaf; evergreen

flowers   showy, solitary on short branchlets; calyx petaloid, yellow to brown, petals absent; flower subtended by 1 or 2 pair of 3-lobed bracts; stamens numerous

fruit   achene (6–8 mm long), solitary in each flower, glabrous, coriaceous

other   bark gray to ashy, becoming black with age, finely striate

HISTORIC, FOOD, AND MEDICINAL USES   none

LIVESTOCK LOSSES   none

FORAGE VALUE   fair for cattle, sheep, goats, and deer during the winter; survives on overgrazed ranges because of the spiny character of the branches

HABITAT   desert mesas and foothills in the pinyon-juniper type, frequently in sandy or gravelly soils

Mexican cliffrose (*Cowania mexicana*)

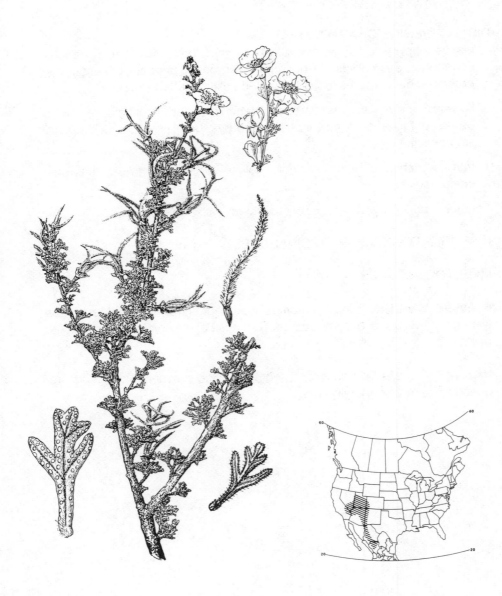

Family	ROSACEAE
Species	*Cowania mexicana* D. Don
Common Name	Mexican cliffrose (quininebush)
Life Span	Perennial
Origin	Native
Season	Evergreen

GROWTH CHARACTERISTICS   shrub or small tree (to 8 m tall), with a rounded crown, flowers in early spring, reproduces by seeds

DISTINGUISHING CHARACTERISTICS

leaves   alternate or clustered, short pedicels with stalked glands, obovate to narrow-spatulate in outline (0.6–1.5 cm long), divided into 3–7 linear lobes, margins enrolled, glandular-punctate, dark green above, paler and hairy below

flowers   single at tips of branches; calyx small, 5-lobed, pubescent; 5 petals (6–8 mm long), cream to yellow in color; numerous stamens; 5–10 pistils, styles elongating in fruit

fruit   achenes (6–8 mm long), leathery, tipped with a feathery style (5 cm long)

other   leaves appear crowded at ends of short branchlets; bark scaly and shreddy with age; glandular-puberulent when young

HISTORIC, FOOD, AND MEDICINAL USES   used by Hopi Indians medicinally as a wash for wounds, wood used for arrow shafts; Indians of the Southwest used wood and fiber for baskets, sandals, ropes, and clothing; cultivated ornamental in rock gardens

LIVESTOCK LOSSES   none

FORAGE VALUE   important browse for cattle and sheep; staple feed for mule deer in some areas

HABITAT   dry soils on slopes and mesas

Apache plume (*Fallugia paradoxa*)

Family	ROSACEAE
Species	*Fallugia paradoxa* (D. Don) Endl.
Common Name	Apache plume
Life Span	Perennial
Origin	Native
Season	Cool

GROWTH CHARACTERISTICS   shrub (0.5–1.5 m tall), straggly, flowers June–August when moisture is available for growth

DISTINGUISHING CHARACTERISTICS

leaves   alternate but fascicled, wedge-shaped in outline (1–2 cm long), divided into 3–7 finger-like lobes, margins entire and rolled inward, underside rust-tomentose, small stipules present

flowers   showy, terminal, solitary or few; calyx fused to form a tube, 5-lobed; petals 5, white, rounded and spreading; stamens numerous

fruit   achene (3 mm long), tipped with elongated feathery styles (2.5–4.0 cm long, often purple or red with age, numerous, thus appearing as feathery balls (resembling on Apache headdress, hence common name)

other   bark shreddy, white when young, turning dark with age

HISTORIC, FOOD, AND MEDICINAL USES   American Indians used bundles of twigs as brooms and older stems for arrow shafts; Hopi Indians made decoction from leaves and used as a hair growth stimulant; used as an ornamental

LIVESTOCK LOSSES   none

FORAGE VALUE   fair for cattle and goats, good winter forage for sheep and big game; used for erosion control

HABITAT   dry deserts or rocky slopes

Shrubby cinquefoil (*Potentilla fruticosa*)

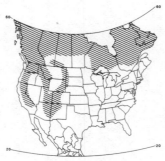

Family	ROSACEAE
Species	*Potentilla fruticosa* L.
Common Name	Shrubby cinquefoil (bush cinquefoil, yellow rose)
Life Span	Perennial
Origin	Native
Season	Cool

GROWTH CHARACTERISTICS   shrub (0.3–1.0 m tall), much-branched, stems leafy, flowers in early summer, reproduces by seeds

DISTINGUISHING CHARACTERISTICS

leaves   alternate, pinnately compound, 3–7 foliate, leaflets linear-oblong (0.5–2.5 cm long), tapering at each end, margins entire and often rolled inward, leaflet surfaces with appressed whitish hairs, especially below, upper 3 often united at base; stipules scarious

flowers   in small loose cymes or solitary; sepals and petals 5, petals a showy yellow, rounded (0.5–1.5 cm long); stamens numerous

fruit   pubescent achenes

other   bark brown to red, shreddy with age

HISTORIC, FOOD, AND MEDICINAL USES   used for erosion control to some extent; ornamental value in some regions; steeped leaves can be used as a tea substitute

LIVESTOCK LOSSES   none

FORAGE VALUE   important sheep and goat browse in the Southwest, occasionally browsed by white-tail deer, browsed extensively by mule deer

HABITAT   subarctic regions, wet alpine meadows to rocky moist grounds at higher elevations

Chokecherry (*Prunus virginiana*)

Family	ROSACEAE
Species	*Prunus virginiana* L.
Common Name	Chokecherry
Life Span	Perennial
Origin	Native
Season	Cool

GROWTH CHARACTERISTICS   large shrub or small tree (to 12 m tall), deciduous, flowers April–July, fruit matures July–September, reproduces from seeds

DISTINGUISHING CHARACTERISTICS

leaves   alternate, simple, thin, oval to oblong (2–10 cm long and 1–5 cm wide), with an abruptly acuminate tip, sharp serrations on margin, dark green above, paler below, veins sometimes pubescent, turn yellow in fall

flowers   in short dense racemes (5–15 cm long), corolla small (4 mm wide), white; sepals obtuse, glandular-pectinate

fruit   drupe (6–8 mm long), globose, dark red to black, juicy, acidulous, on pedicels

other   twigs reddish-brown, slender; bark gray to black

HISTORIC, FOOD, AND MEDICINAL USES   fruits are used for jams; American Indians used bark extract to cure diarrhea; fruits to treat canker sores and add to pemmican; and wood for arrows, bows, pipe stems

LIVESTOCK LOSSES   poisonous, contains toxic quantities of hydrocyanic acid (HCN) in leaves, poisonous to all classes of livestock, poisoning generally occurs when other forage is not available, or after drought or freezing

FORAGE VALUE   poor to fair for cattle and sheep, good for wildlife browse; fruit is important food source for grouse, quail, prairie chickens, and several songbirds

HABITAT   moist soils in open sites near fence rows, roadsides, and windbreak borders; hillsides and canyons

Antelope bitterbrush (*Purshia tridentata*)

Family	ROSACEAE
Species	*Purshia tridentata* (Pursh) DC.
Common Name	Antelope bitterbrush (bitterbrush)
Life Span	Perennial
Origin	Native
Season	Evergreen

GROWTH CHARACTERISTICS   shrub, prostrate to erect (0.6–3.0 m tall), evergreen to late-deciduous, flowers April–August, fruit is present July–September, excellent drought resistance

DISTINGUISHING CHARACTERISTICS

leaves   alternate but crowded so appear fascicled, wedge-shaped (0.5–2.5 cm long), apex tridentate, margins rolled inward, hairy to glabrous and dark green above, densely white-woolly beneath, stipules small

flowers   solitary; calyx tube funnelform, pubescent to glandular, 5-lobed; petals 5 (5–8 mm long), yellow; stamens 20–25; style beak-like and persistent

fruit   pubescent achene (0.8–1.2 cm long)

other   stems gray to brown, pubescent at first then becoming glabrous; buds small and scaly

HISTORIC, FOOD, AND MEDICINAL USES   cultivated as an ornamental

LIVESTOCK LOSSES   none

FORAGE VALUE   important cattle and sheep browse; excellent for wildlife including antelope, deer, pica, squirrels, and chipmunks

HABITAT   hillsides and slopes, plains, open areas in forests

Wild rose (*Rosa woodsii*)

Family	ROSACEAE
Species	*Rosa woodsii* Lindl.
Common Name	Wild rose (Woods rose, rosa silvestre)
Life Span	Perennial
Origin	Native
Season	Cool

GROWTH CHARACTERISTICS   shrub (1 m tall), deciduous, growth starts in early spring, flowers June—July, reproduces by seeds

DISTINGUISHING CHARACTERISTICS

leaves   alternate (2—6 cm long), pinnately compound, 5—9 leaflets, elliptic, coarsely or sharply serrate, glabrous and shiny above; prominent stipules united at bases (4—7 mm wide), glandular-pubescent on back

flowers   pink to red (3 cm wide), solitary or in a corymb; sepals often glandular (1.0—1.5 cm long), erect or spreading in fruit; stamens numerous

fruit   hypanthium forming a fleshy rounded red hip, fruit an achene

other   bark reddish-brown to gray; stems with straight or recurved prickles

HISTORIC, FOOD, AND MEDICINAL USES   Europeans utilized hips as a source for vitamin A and C (pills manufactured), rose hip powder used as a flavoring in soups and for making syrup; American Indians utilized the plant in several ways, young shoots as a potherb, leaves steeped for tea, petals eaten raw, in salads, candied or made into syrup, inner bark smoked like tobacco, and dried petals were stored for perfume

LIVESTOCK LOSSES   prickles (thorns) may injure soft tissue

FORAGE VALUE   fair to good sheep and cattle browse; good wildlife browse for elk and deer; small mammals and birds feed on hips

HABITAT   plateaus and dry slopes, prairies and thickets

Quaking aspen (*Populus tremuloides*)

Notes:

Family	SALICACEAE
Species	*Populus tremuloides* Michx.
Common Name	Quaking aspen (trembling aspen, aspen, chopo)
Life Span	Perennial
Origin	Native
Season	Cool

GROWTH CHARACTERISTICS   tree (6–30 m tall), deciduous, rounded top, grows rapidly from root sprouts, seldom reproduces by seeds, short-lived

DISTINGUISHING CHARACTERISTICS

leaves   simple, alternate, nearly round to broadly ovate with a point at the apex, margins finely serrated, dark green shiny above, pale green beneath, main vein conspicuous and white, weak petiole (4.0–6.5 cm long), slender and flattened, leaves change bright yellow or yellow-orange in fall

flowers   on drooping catkins (3.0–6.4 cm long), apetalous; scales of 3–5 hairy-fringed lobes

fruit   capsule, light green to brown (3–5 mm long)

other   twigs slender reddish-brown to gray, shiny; buds resinous, reddish-brown, shiny; bark green to gray or white, smooth

HISTORIC, FOOD, AND MEDICINAL USES   bark used by pioneers and American Indians as a fever remedy and for scurvy, contains salicin (similar to the main ingredient in aspirin) from which the family name SALICACEAE was derived; wood is used for pulp and lumber

LIVESTOCK LOSSES   none

FORAGE VALUE   fair to good for sheep, fair for cattle; twigs, bark, and buds are browsed by beavers, pika, deer, elk, moose, bears, squirrels, cottontail and snowshoe rabbits, and porcupine; seeds eaten by grouse and other birds

HABITAT   diverse soil types; shallow rocky, or clayey to rich sandy soil, common throughout forested regions and parklands

413

Wax currant (*Ribes cereum*)

414

Family	SAXIFRAGACEAE
Species	Ribes cereum Dougl.
Common Name	Wax currant (squaw currant)
Life Span	Perennial
Origin	Native
Season	Cool

GROWTH CHARACTERISTICS   shrub (0.1–1.2 m tall), much-branched, flowers June–July

DISTINGUISHING CHARACTERISTICS

leaves   alternate and fascicled, petiolate, kidney-shaped (1–4 cm wide, broader than long), obscurely 3–5 lobed, scallop-toothed margins, upper surface subglabrous to shiny, lower surface hairy and white to waxy gland-dotted (upper surface may also be)

flowers   floral tube (6–8 mm long), greenish-white to pink; sepals rounded (1.0–1.5 mm)

fruit   berry, red, slightly glandular-hairy

other   fragrant; bark white

HISTORIC, FOOD, AND MEDICINAL USES   edible berries; American Indians ate fresh berries or dried to preserve, these were later eaten with raw mutton and deer fat

LIVESTOCK LOSSES   none

FORAGE VALUE   fair to good for wildlife, poor to fair for livestock, but is often abundant, therefore a rather important browse plant

HABITAT   dry, rocky, open slopes; hills and ridges

Blue penstemon (*Penstemon glaber*)

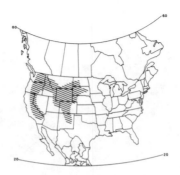

416

Family	SCROPHULARIACEAE
Species	*Penstemon glaber* Pursh
Common Name	Blue penstemon (smooth beardtongue, sawsepal penstemon)
Life Span	Perennial
Origin	Native
Season	Cool

GROWTH CHARACTERISTICS   forb (30–60 cm tall), erect, roots woody and fibrous, reproduces by seeds

DISTINGUISHING CHARACTERISTICS

leaves   opposite, oblong to lanceolate (5–10 cm long), sessile (basal leaves narrowed to petioles), margin entire, glaucous

flowers   irregular calyx 5-parted with scarious margins (5 mm long); corolla bluish-purple (3 cm long), somewhat 2-lipped with rounded lobes; sterile filament bearded

fruit   2-valved capsule, ovoid (1 cm long)

other   woody roots and root crowns

HISTORIC, FOOD, AND MEDICINAL USES   some American Indians used a wet dressing of *Penstemon* spp. to treat snakebite, others drank a penstemon tea to stop vomiting; also grown as an ornamental

LIVESTOCK LOSSES   may accumulate selenium

FORAGE VALUE   fair for mule deer and sheep, worthless for cattle

HABITAT   plains and hills

417

Spiny hackberry (*Celtis pallida*)

Notes:

Family	ULMACEAE
Species	*Celtis pallida* Torr.
Common Name	Spiny hackberry (desert hackberry, granjeno)
Life Span	Perennial
Origin	Native
Season	Warm

GROWTH CHARACTERISTICS   shrub (to 5.5 m tall), forms dense thickets

DISTINGUISHING CHARACTERISTICS

leaves   simple (to 3 cm long and 2 cm wide), alternate, ovate to ovate-oblong (sometimes elliptic), entire or slightly crenate-dentate, pubescent to slightly scabrous, unequal at base, 3 prominent veins (2 glands at vein junctions on underside of leaf)

flowers   small, white in cymes of 3–5 flowers (in elongated clusters at the base of younger leaves)

fruit   drupe (6 mm long), edible, ovoid, glabrous, orange-yellow and red

other   numerous spreading white-puberulent branches; spines (to 2.5 cm) in pairs form a V-shape (not opposite); main branch has a zigzag appearance

HISTORIC, FOOD, AND MEDICINAL USES   wood for fence posts and firewood; fruits ground and eaten by Southwestern Indians; good honey plant

LIVESTOCK LOSSES   spines may cause minor injuries

FORAGE VALUE   no forage value for domestic livestock; fruits eaten by deer, raccoons, rabbits, quail, small birds; browsed only under overgrazed conditions; valuable erosion control

HABITAT   mesas, foothills, thickets, and brushlands

Creosotebush (*Larrea tridentata*)

Notes:

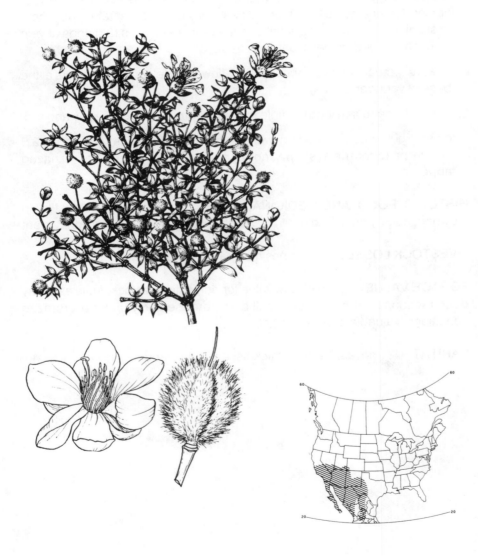

Family	ZYGOPHYLLACEAE
Species	*Larrea tridentata* (DC.) Cov.
Common Name	Creosotebush (gobernadora, hediondilla)
Life Span	Perennial
Origin	Native
Season	Evergreen

GROWTH CHARACTERISTICS   shrub (to 3 m tall), no well-defined trunk, numerous slender stems arise from near ground level; often in pure stands, xerophyte, important soil protector and stabilizer

DISTINGUISHING CHARACTERISTICS

leaves   opposite, subsessile to short-petiolate, leaflets 2, opposite, fused at the base, ovate to oblong or obovate, leathery, glossy dark green to yellow-green, resinous (0.5–1.0 cm long), brown stipules at base of petiole glandular hairy

flowers   solitary, petals 5, bright yellow (partly twisted, like vanes on a windmill)

fruit   capsule, spherical, densely covered with white- to red-woolly long hairs tipped by a thread-like stalk

other   bark gray to black; nodes dark, conspicuous (gives stem a "bumpy" appearance); stems resinous causing plant to have a creosote-like odor especially when moist or burned

HISTORIC, FOOD, AND MEDICINAL USES   American Indians used decoctions as antiseptic, medicine, and fuel; Pima Indians used scale insect larvae found on the plant for mending pottery and cementing arrowheads

LIVESTOCK LOSSES   none

FORAGE VALUE   worthless to livestock

HABITAT   alluvial plains, sandy and gravelly soils of mesas and hillsides

Glossary

A   Prefix meaning without

Abaxial   On the side away from the axis

Acaulescent   Stemless, without an aboveground stem or apparently so

Achene   A 1-seeded, indehiscent fruit with a relatively thin wall in which the seed coat is not fused to the ovary wall

Acuminate   Gradually tapering to a sharp point; compare with acute

Acute   Sharp-pointed, but less tapering than acuminate

Adaxial   On the side nearest the axis

Adnate   Attached or grown together, fusion of unlike parts, such as palea and caryopsis in the genus *Bromus*

Alternate   Located singly at each node

Ament   A dense spike or raceme with many small, usually naked, flowers. A catkin such as in *Populus*

Annual   Within 1 year; applied to plants which do not live more than 1 year

Anterior   In front of; in a flower, the side away from the axis and adjacent to the bract

Antrorse   Directed upwards or forwards; opposed to retrorse

Apomixis   Process of setting seed without fertilization

Appressed   Lying against an organ, flatly pressed against

Arcuate   Curved like a bow

Aristate   Awned or tapering; a very long, narrow apex

Aromatic   Fragrant or having an odor; bearing essential oils

Articulate   Jointed, provided with nodes; separating clearly at maturity

Ascending   Growing or curving upward; obliquely upward

Attenuate   Gradually narrowed to a slender apex or base

Auricle   Applied to ear-like lobes at the base of leaf blades

Awl-shaped   Narrow and sharp pointed, gradually tapering from a narrow base to a pointed apex

Awn   A slender bristle at the end, on the back, or on the edge of an organ. In grasses, the extension of a nerve beyond the leaf-like tissue

Axil   Angle between an organ and its axis

Axillary   Growing in an axil

Banner   Upper petal (standard) of the papilionaceous flower in the FABACEAE

Barbed   furnished with retrorse projections

Beak   A hard point or projection

Bearded   Furnished with long, stiff hairs

Bifid   1-cleft or 2-lobed; applied to the summit of glumes, lemmas, paleas, petals, or leaflets

Bipinnate   Twice pinnate

Blade   The part of the leaf above the sheath, petiole, or petiolule

Bract   Reduced leaves

Branch   A lateral stem

Bristle   A stiff, slender appendage; in grasses, a reduced branch such as in *Setaria*

Bulbous   Swollen at the base, like a bulb or corm

Caespitose   Tufted, several or many stems in a close tuft, such as bunch grass

Callus   The indurate downward extension of tissue from the mature lemma in *Stipa, Aristida,* and some other genera

Calyx   The sepals of a flower considered collectively, usually green bracts

Campanulate   Shaped like a bell

Canescent   Pale or gray colored because of a dense, fine pubescence

Capillary   Fine and slender, hair-like such as a branch or awn

Capsule   A dry, dehiscent fruit of more than 1 carpel with more than 2 seeds

Carpel   The modified leaf forming the ovary, may be compound or simple

Catkin   A dense spike or raceme with many small, usually naked, flowers; an ament such as in willows or aspen

Caudex   A short, usually woody, vertical stem located just below the soil surface; new shoots arise from the caudex each year

Cauline   Pertaining to the stem or belonging to the stem

Channeled   Deeply grooved

Chartaceous   Having the texture of writing paper and usually not green

Ciliate   Fringed with hairs on the margin

Cleistogamous   Applied to flowers or florets fertilized without opening, such as in *Leptochloa* and *Stipa* species; closed marriage

Collar   The area on the outer side of a leaf at the junction of the blade and sheath

Compressed   Flattened laterally

Cones   A cluster of scales on an axis, scales may be persistent or deciduous

Connate   Fusion of like parts, such as petals to form a corolla tube

Contracted Inflorescences that are narrow or dense, frequently spike-like

Convex Rounded on the surface

Convolute Rolled longitudinally

Coriaceous Leathery in texture

Corolla All of the petals considered collectively

Corymb A simple racemose inflorescence that is flat-topped; an indeterminate inflorescence

Corymbiform Having the form, but not necessarily the structure of a corymb

Cotyledon A leaf of the embryo of a seed; the seed leaf

Crenate Having rounded teeth; scalloped margins

Crested With an elevated ridge or appendage on the top or back

Crown Persistent base of a herbaceous perennial

Culm The jointed stem of a grass, hollow or solid

Cyme A convex or flat-topped flower cluster with the central flower the first to open; a determinate inflorescence

Cymose Resembling a cyme or bearing cymes

Deciduous Not persistent, but falling away in less than one year

Decumbent Curved upward from a horizontal or inclined base

Dehiscent Opening at maturity along a definite suture

Deltoid Triangular, shaped like the Greek letter delta

Dentate With pointed, coarse teeth spreading at right angles to the margin

Denticulate Diminutive of dentate

Diffuse Open and much-branched, loosely branching

Digitate Several members arising from the summit of a support; like the fingers arising from the hand as a point of origin

Dioecious Unisexual flowers on separate plants, pistillate and staminate flowers on separate plants

Disarticulating Separating at maturity at a node or joint

Discoid Resembling a disk; in the ASTERACEAE, with all the flowers of a head tubular and perfect

Disk An outgrowth of the receptacle that surrounds the base of the ovary or ovaries

Dissected Deeply divided into numerous parts

Distichous Conspicuously 2-ranked leaves, leaflets or flowers

Divaricate Widely and stiffly divergent

Divergent Widely spreading

Dorsal Relating to the back of an organ; opposite the ventral side

Drupe A fleshy fruit; indehiscent, usually with a single seed (e.g., a cherry), with a stony endocarp

Elliptic Shaped like an ellipse; narrowly pointed at the ends and widest in the middle

Elongate Narrow, the length many times the width or thickness

Erose   Irregularly notched at the apex; appearing gnawed or eroded

Exserted   Protruding or projecting beyond; not included

Fascicle   A small bundle or cluster, such as pine needles in clusters of 2 to 5

Fertile   Capable of producing fruit; especially used in grasses; does not refer to stamen presence or absence

Filiform   Thread-like, long and very slender

Flexuous   Bent alternately in opposite directions; a wavy form

Floccose   Covered with long, soft, fine hairs that are loosely spreading, these hairs rub off easily

Floret   Lemma and palea with included flower of POACEAE; also disk flowers of ASTERACEAE

Floriferous   Flower-bearing

Fruit   Ripened ovary (pistil); the seed bearing organ

Fusiform   Spindle-shaped

Geniculate   Bent abruptly, like a knee; awns or plant bases are bent in this manner

Glabrate   Nearly glabrous or becoming so with age

Glabrous   Without hairs

Gland   A protuberance or depression, that appears to secrete a fluid

Glandular   Supplied with glands

Glaucous   Covered with a waxy coating that gives a blue-green color

Globose   Nearly spherical in shape

Glumes   The pair of bracts at the base of a spikelet in grasses

Glutinous   With a firm, sticky substance covering the surface

Herbaceous   Not woody, dying each year, or dying back to the crown

Hirsute   With straight, rather stiff hairs

Hispid   With stiff or rigid hairs; bristly hairs

Hyaline   Thin and translucent or transparent

Hypanthium   A ring or cup around the ovary formed by a fusion of the bases of sepals, stamens, and petals; a modified receptacle

Imbricate   Overlapping; like shingles on a roof

Indehiscent   Not opening, staying closed at maturity; not splitting

Indurate   Hard

Inflated   Puffed up, bladdery

Inflexed   Turned in at the margins

Inflorescence   The flowering part of a plant; mode of flowering

Internerves   Spaces between the nerves

Internode   The part of a stem between 2 successive nodes

Involucre   A whorl or circle of bracts below the flower or spikelet cluster

Involute   Rolled inward from the edges, the upper surface within

Keel   The sharp fold or ridge at the back of a compressed sheath, blade, glume, lemma, or palea of POACEAE or the united lower petals of FABACEAE

Lacerate   Appearing torn at the edge or irregularly cleft
Lacinate   Deeply cut into narrow segments
Lanate   Woolly with long intertwined curly hairs
Lancelinear   Shaped like a narrow lance
Lanceolate   Rather narrow, tapering to both ends, widest below the middle
Leaf   Lateral organ of a stem
Legume   FABACEAE fruit composed of a single carpel, but with 2 sutures and dehiscing at maturity along the sutures
Lemma   Abaxial bract of the floret that subtends the grass flower and palea
Ligulate   In the ASTERACEAE, referring to flowering heads solely composed of the flat, strap-shaped flowers on the margin (ray flowers) of the disk, or flowers of the head that are strap-like
Ligule   The appendage, membrane, or ring of hairs on the adaxial side of a leaf at the junction of the sheath and blade
Linear   Long and narrow with parallel sides
Lobe   The projecting part of an organ with divisions less than ½ the distance to the base or midvein, usually rounded or obtuse

Membranous   Thin; opaque, not green; like a membrane
-merous   Referring to number of parts
Monoecious   Plants with male and female flowers at different locations on the same plant with all flowers unisexual
Mottled   Marked with spots or blotches
Mucronate   Tipped with a short, slender, sharp point or awn

Nerve   The vascular bundles or veins of the leaf blades, culms, glumes, paleas, and lemmas or other organs
Nodding   Inclined somewhat from the vertical
Nutlet   A small, usually 1-seeded, hard fruit that is indehiscent; a small nut

Ob   A prefix meaning inversely
Obovate   Opposite of ovate with the widest part towards the far end; egg shaped with the widest part above the middle
Obtuse   Shape of an apex, with an angle greater than 90°
Orbicular   Nearly circular in outline
Oval   Broadly elliptic
Ovate   Shaped like a hen's egg with the broadest portion towards the base
Ovule   Not yet mature seed located in the ovary; the egg containing part of the ovary

Palea   The adaxial bract of a floret; the upper bract subtending the flower

Palmate   With 3 or more lobes, nerves, or leaflets arising from a common point

Panicle   Inflorescence with a main axis and rebranched branches

Paniculiform   Having the shape of a panicle

Papery   Having the texture of writing paper

Papilionaceous   A flower type in the FABACEAE having a banner petal, 2 wing petals, and 2 partially fused to fused keel petals

Pappus   A group of hairs, scales, or bristles that crown the summit of the achene in the ASTERACEAE flower

Pectinate   Comb-like, divided into numerous narrow segments

Pedicel   The stalk of a spikelet or single flower in an inflorescence

Pedicellate   Having a pedicel

Penduncle   The stalk of a flower cluster or spikelet cluster

Pendulous   Suspended or hanging downward, drooping

Perennial   Lasting more than 2 years; applied to plants or plant parts which live more than 2 years

Perfect   Applied to flowers having both stamen and pistil

Pericarp   The fruit wall; wall of a ripened ovary

Perigynium   An inflated sac that encloses the achene of *Carex*

Petal   A part or member of the corolla, usually brightly colored

Petaloid   Petal-like

Petiolate   With a petiole

Petiole   The stalk of a leaf blade

Phyllary   A bract of the involucre at the outside of the head of flowers in the ASTERACEAE

Pilose   With long soft, straight hairs

Pinnate   Having 2 rows of lateral divisions along a main axis (like barbs of a feather)

Pinnatifid   Deeply cut in a pinnate manner, but not cut entirely to the main axis

Pistillate   Applied to flowers bearing pistils only; unisexual flowers

Plumose   Feathery, with long pubescence or pinnately arranged bristles

Primary branch   Branch arising directly from the main inflorescence axis

Puberulent   Diminutive of pubescent

Pubescent   Covered with short, soft hairs

Punctate   Having dots, usually with small glandular pits

Pungent   Sharp and penetrating odor, or a rough surface that is uncomfortable to touch because of firm, or sharp points

Pustulate   Having small eruptions or blisters

Pyramidal   Pyramid-shaped

Raceme   An inflorescence in which the spikelets or flowers are pediceled on a rachis

Racemose   Raceme-like inflorescences

Rachilla   A small axis applied especially to the axis of a spikelet

Rachis   The axis of a spike or raceme inflorescence or pinnately compound leaf

Radiate   Term used to describe the ASTERACEAE flower arrangement with the marginal flowers ligulate and the disk or central flowers tubular; spreading from a common center

Radiating   Spreading from a common center

Ray   Ligulate flowers in ASTERACEAE

Receptacle   The upper end of the stem of a plant to which the flowering parts are attached

Resinous   Producing any of numerous viscous substances such as resin or amber

Reticulate   In the form of a network, netted as many leaf veins in dicots

Retrorse   Pointing backward toward the base

Revolute   Rolled under along the margin toward the abaxial surface

Rhizome   An underground stem with nodes, scale-like leaves and internodes

Rhizomatous   Having rhizomes

Rosette   A basal, usually crowded, whorl of leaves

Rudiment   Imperfectly developed organ or part, usually non-functional

Rudimentary   Under-developed

Rugose   Wrinkled surface

Sagittate   Triangular or arrowhead-shaped with the lobes turned downward. See *Balsamorhiza* leaf bases

Scabrous   Rough to the touch; short angled hairs requiring magnification for observation

Scale   Reduced leaves at the base of a shoot or a rhizome

Scape   A leafless peduncle arising from the ground or basal whorl of leaves and bearing 1 or more flowers. See *Taraxacum*

Scapose   Bearing a flower or flowers on a scape or resembling a scape

Scurfy   Covered with minute scales or specialized mealy hairs

Sepal   A member of the calyx bracts, usually green

Serrate   Saw-toothed margins, with teeth pointing toward the apex

Serrulate   Minutely serrate

Sessile   Without a pedicel or stalk

Setaceous   Bristle-like hairs

Sheath   The lower part of a leaf that encloses the stem

Silique   A long, slender capsular fruit of 2 carpels, a type of capsule in the BRASSICACEAE

Sinus   Identation between 2 lobes or segments

Spathe   A modified sheathing bract of the inflorescence

Spatulate   Shaped like a spatula, being broader above than below

Spike   An unbranched inflorescence in which the spikelets or flowers are sessile on a rachis

Spikelet   The unit of inflorescence in grasses usually consisting of 2 glumes, 1 or more florets, and a rachilla

Spinescent   Tipped by spines or thorns

Spur   Any slender, hollow projection of a flower. See *Delphinium* flowers

Staminate   Containing stamens only in the flower, unisexual flowers

Stellate   Star-shaped, usually referring to hairs with many branches from the base

Sterile   Without functional pistils, may or may not bear stamen

Stipe   In general, a stalk or stem that supports an organ

Stipules   Appendages, usually leaf-like, occurring in pairs, 1 on either side of the petiole base, may be modified to spines

Stolon   A horizontal, above ground, modifided propagating stem with nodes, leaves, and internodes

Stoloniferous   Bearing stolons

Striate   Marked with slender longitudinal grooves or lines; appearing striped

Strict   Narrow, with close, upright branches

Strigose   Rough with short, stiff hairs or bristles

Sub   A prefix to denote somewhat, slightly, or in less degree

Subulate   Awl-shaped

Suffrutescent   Having a woody or shrubby base

Tawny   Pale brown or dirty yellow

Teeth   Pointed lobes or divisions

Terete   Cylindric and slender

Terminal   Borne at or belonging to the extremity or summit

Thyrse   A flower cluster of racemosely arranged cymes organized into an elongate panicle

Tomentose   A surface covered with matted and tangled hairs

Tridentate   3-toothed, such as big sagebrush leaves

Trifoliate   Having 3 leaflets. See *Medicago*

Truncate   Ending abruptly; appearing to be cut off at the end

Tubercle   A small projection from the surface of an organ or structure

Tuberculate   Furnished with small projections

Tuft   Cluster

Ultimate   Smallest subdivisions

Undulate   Strong wavy in a perpendicular plane

Unisexual   Said of flowers containing only stamens or only pistils

Urceolate   Shaped like an urn

Utricle   A small 1-seeded fruit with a thin wall, dehising by the breakdown of the thin wall

Valve   1 portion of a compound ovary, part of a pod or capsule

Villous   With long, soft macrohairs; similar to pilose, but with a higher density of hairs

Viscid   Sticky or clammy

Whorl   A cluster of several branches or leaves around the axis

Wing   A thin projection or border

Xerophyte   A plant adapted to a dry habitat

Authorities

Ait.   William Aiton (1731–1793), English botanist

A. Nels.   Aven Nelson (1859–1952), professor of botany and president of the University of Wyoming

Asch. & Schweinf.   Paul Friedrich August Ascherson (1834–1913), German botanist and Georg August Schweinfurth (1836–1925), German botanist

Beal   William James Beal (1833–1924), Michigan State University agrostologist

Beauv.   Ambroise Marie Francois Joseph Palisot de Beauvois (1752–1820), French naturalist

Benth.   George Bentham (1800–1884), English taxonomist

Bert.   Antonio Bertoloni (1775–1869), Italian professor of botany

Bieb.   Baron Friedrich August Marschall von Bieberstein (1768–1826), German explorer (Russia and the Caucasus)

Bisch.   Gottlieb Wilhelm Bischoff (1797–1854), German professor

Blake   Sidney Fay Blake (1892–1959), United States Department of Agriculture scientist

Boland.   Henry Bolander (1831–1897), American botanist

Boott   Francis Boott (1805–1887), American botanist and authority on *Carex*

Britt.   Nathaniel Lord Britton (1859–1934), director-in-chief of the New York Botanical Garden

Britt. & Rusby   N. L. Britton (1859–1934) and Henry Hurd Rusby (1855–1940), Dean of the New York College of Pharmacy and active collector, especially in South America

B.S.P.   N. L. Britton (1859–1934), Emerson Ellick Sterns (1846–1926), and Justus Ferdinand Poggenburg (1840–1893); American botanists

Buckl.   Samuel Botsford Buckley (1809–1884), naturalist and state geologist of Texas

Bush   Benjamin Franklin Bush (1858–1937), postmaster at Independence, Missouri, and amateur botanist

C. A. Mey. Carl Anton von Meyer (1795–1855), director of the St. Petersburg, Russia, botanical garden

Cav. Antonio Jose Cavanilles (1745–1804), Spanish botanist

Clayton John Clayton (1685–1773), physician in Virginia and amateur botanist

Coult. & Rose John Merle Coulter (1851–1928), professor of botany, University of Chicago, and Joseph Nelson Rose (1862–1928), American botanist

Cov. Fredrick Vernon Coville (1867–1937), curator of the United States National Herbarium

DC. Augustin Pyramus de Candolle (1778–1841), Swiss botanist and professor of botany

D. Dietr. David Nathanael Friedrich Dietrich (1799–1888), German botanist

D. Don David Don (1799–1841), English botanist

Dougl. David Douglas (1798–1834), Scotch botanical collector in northwestern America

Dun. Michel Felix Dunal (1789–1856), French botanist

Eat. Amos Eaton (1776–1842), American botanist, produced first botanical manual in America with descriptions in English

Elmer Adolph Daniel Edward Elmer (1870–1942), American botanist

Endl. Stephan Friedrich Ladislaus Endlicher (1804–1849), Austrian botanist

Engelm. George Engelmann (1809–1884), physician and botanist in St. Louis, Missouri

Fisch. & Trautv. Friedrich Ernst Ludwig von Fischer (1782–1854), director of St. Petersburg, Russia, botanical garden and Ernst Rudolph von Trautvetter (1809–1889), Russian botanist

Fourn. Eugene Pierre Nicolas Fournier (1834–1884), physician in Paris and amateur botanist

Gaertn. Joseph Gaertner (1732–1791), German botanist

Geyer Carl Andreas Geyer (1809–1853), Austrian botanist that collected in northwestern United States

Gray Asa Gray (1810–1888), professor of botany at Harvard University

Greene Edward Lee Greene (1842–1915), American botanist at the University of California

H. & A. Sir William Jackson Hooker (1785–1865), English botanist and director of the Royal Botanic Gardens at Kew, and George Arnold Walker Arnott (1799–1869), Scotch botanist

Harv. & Gray William Henry Harvey (1811–1866), Irish botanist and Asa Gray (1810–1888), professor of Botany at Harvard University

H.B.K. Baron F.W.H.A. von Humboldt (1769–1859), A.J.A. Bonpland (1773–1858), Carl Kunth (1788–1850), conducted and described a scientific expedition to tropical America

Henr.   Jan Theodoor Henrard (1881–?), conservator Rijksherbarium, Netherlands

Hitchc.   Albert Spear Hitchcock (1865–1935), American agrostologist

Hook.   Sir William Jackson Hooker (1785–1865), director of the Royal Botanic Gardens at Kew, England

Host   Nicolaus Thomas Host (1761–1834), Austrian physician and botanist ꙮ

Jacq.   Nikolaus Joseph Baron von Jacquin (1727–1817), Austrian taxonomist

J. G. Smith   Jared Gage Smith (1866–1925), American agrostologist

J. T. Howell   John Thomas Howell (1903–   ), California botanist

Kell.   Albert Kellogg (1813–1887), physician and botanist from San Francisco

K. Schum.   Karl Moritz Schumann (1851–1904), German botanist

Kunth   Carl Sigismund Kunth (1788–1850), German botanist

L.   Carolus Linnaeus (1707–1778), Swedish botanist and author of *Species Plantarum* upon which botanical nomenclature is based

Lag.   Mariano Lagasca y Segura (1776–1839), Spanish professor

Lam.   Jean Baptiste Antoine Pierre Monnet de Lamarck (1744–1829), French botanist

Laws.   Peter Lawson (?–1820), and Sir Charles Lawson (1794–1873), the son (?), Scotch nurserymen

Leyss.   Freidrich Wilhelm von Leysser (1731–1815), German botanist and author of the Flora of Halle

L'Her.   Charles Louis L'Heritier de Brutelle (1746–1800), French magistrate and botanist

Lindl.   John Lindley (1799–1865), English professor of botany

Link   Johann Heinrich Friedrich Link (1767–1851), German professor

Macoun   John Macoun (1831–1920), Irish-born, Canadian botanist

Malte   Malte Oscar Malte (1880–1933), chief botanist of the National Herbarium of Canada

Merr.   Elmer Drew Merrill (1876–1956), American botanist at Harvard University

Michx.   Andre Michaux (1746–1802), French botanist and explorer of North America

Moench   Conrad Moench (1744–1805), German botanist

Moq.   Christian Horace Benedict Alfred Moquin-Tandon (1804–1863), French botanist

Muenchh.   Otto Frieherr von Muenchhausen (1716–1774), German botanist

Muhl.   Gotthilf Heinrich Ernest Muhlenberg (1753–1815), German-educated Lutheran minister and pioneer botanist in Pennsylvania

Nash   George Valentine Nash (1864–1921), agrostologist and head gardener at the New York Botanical Garden

Nees   Christian Gottfried Daniel Nees von Esenbeck (1776–1858), German botanist

Nevski   Sergei Arsenjevic Nevski (1908–1938), Russian agrostologist

Nutt.   Thomas Nuttall (1786–1859), English-American naturalist who collected in western America

Pall.   Peter Simon Pallas (1741–1811), German botanist

Pau

Payne   Willard William Payne (1934–   ), botanist at the University of Illinois

Pers.   Christiaan Hendrick Persoon (1761–1836), South African-French botanist

Phil.   Rudolf Amandus Philippi (1808–1904), Chilean botanist

Piper   Charles Vancouver Piper (1867–1926), American professor in Washington

Poir.   Jean Louis Marie Poiret (1755–1834), French botanist

Port. & Coult.   Thomas Conrad Porter (1822–1901), professor of botany in Pennsylvania and John Merle Coulter (1851–1928), professor of botany at the University of Chicago

Pursh   Fredrick Traugott Pursh (1774–1820), German author and botanical collector in North America

Raf.   Constantine Samuel Rafinesque (1783–1840), pioneer naturalist in Kentucky

Richt.   Karl Richter (1855–1891)

Ricker   Percy Leroy Ricker (1878–?), agronomist with the United States Department of Agriculture

Roth   Albrecht Wilhelm Roth (1757–1834), German physician and botanist

R. & S.   Johann Jacob Roemer (1763–1819), Swiss botanist and Joseph August Schultes (1773–1831), Austrian botanist

Rydb.   Per Axel Rydberg (1860–1931), Swedish-born, American botanist

Sarg.   Charles Sprague Sargent (1841–1927), botanist at Harvard University

Schrad.   Heinrich Adolph Schrader (1767–1836), German botanist and professor

Schult.   Joseph August Schultes (1773–1831), Austrian botanist and professor

Scribn.   Frank Lamson Scribner (1851–1938), agrostologist with the United States Department of Agriculture

Scribn. & Merr.   Frank Lamson Scribner (1851–1938), agrostologist with the United States Department of Agriculture and Elmer Drew Merrill (1876–1956), American botanist at Harvard University

Sennen

Shinners   Lloyd Herbert Shinners (1918–1971), Canadian-born botanist, professor of botany at Southern Methodist University

Simonkai

Sims   John Sims (1749–1831), English botanical editor

Spreng.   Curt Polykarp Joachim Sprengel (1766–1833), professor of botany at Halle, Germany

Steud.   Ernst Gottlieb Steudel (1783–1856), German physician and authority on grasses

Sw.   Olof Peter Swartz (1760–1818), Swedish botanist

Thurb.   George Thurber (1821–1890), American botanist with Mexican Boundary Survey

Torr.   John Torrey (1796–1873), American physician and professor of botany

Torr. & Frem.   John Torrey (1796–1873), American physician and botanist and John Charles Fremont (1813–1890), soldier, explorer, and presidential candidate

Torr. & Gray   John Torrey (1796–1873), American physician and botanist and Asa Gray (1810–1888), professor of botany at Harvard University

Trel.   William Trelease (1857–1945), professor of botany at the University of Illinois

Trin.   Carl Bernhard von Trinius (1778–1844), Russian physician, poet, and authority on grasses

Trin. & Rupr.   Carl Bernhard von Trinius (1778–1844), Russian physician and agrostologist and Franz Josef Ruprecht (1814–1870), Czech-born, Russian botanist

Vasey   George Vasey (1822–1893), English-born, American botanist and curator of the United States National Herbarium

Vent.   Etienne Pierre Ventenat (1757–1808), French botany professor

Vitman   Fulgencio Vitman (1728–1806), Italian botanist

Walt.   Thomas Walter (1740–1789), British-American botanist

Wang.   Friedrich Adam Julius von Wangenheim (1749–1800), German forester

Wats.   Sereno Watson (1826–1892), botanist and assistant to Asa Gray

Weber   Frederick Albert Weber (1830–1903), French botanist and member of French expedition in Mexico

Willd.   Carl Ludwig Willdenow (1765–1812), German botanist

Wood   Alphonso Wood (1810–1881), author of first American book to employ dichotomous keys

Yates   Harris Oliver Yates (1934–   ), American taxonomist

Selected Synonyms

Acacia rigidula Benth.
    SYN = *A. amentacea* DC.
Achillea millefolium L.
    SYN = *A. lanulosa* Nutt.
Agropyron spicatum (Pursh) Scribn.
    SYN = *A. inerme* (Scribn. & Smith) Rydb.
Agropyron trachycaulum (Link) Malte
    SYN = *A. subsecundum* (Link) Hitchc.
            A. pauciflorum Hitchc.
Agrostis exarata Trin.
    SYN = *A. exarata* var. *monolepis* (Torr.) Hitchc.
Agrostis stolonifera L.
    SYN = *A. alba* (of American authors)
Ambrosia deltoidea (Torr.) Payne
    SYN = *Franseria d.* Torr.
Ambrosia dumosa (Gray) Payne
    SYN = *Franseria d.* Gray
Andropogon gerardii Vitman
    SYN = *A. furcatus* Muhl.
            A. hallii Hack.
Andropogon virginicus L.
    SYN = *A. perangustatus* Nash
Arundinaria gigantea (Walt.) Muhl.
    SYN = *A. tecta (Walt.) McClure*
Artemisia nova A. Nels.
    SYN = *A. arbuscula* Nutt.
Atriplex gardneri (Moq.) D. Dietr.
    SYN = in part: *A. nuttallii* Wats.
Bothriochloa saccharoides (Sw.) Rydb.
    SYN = *Andropogon s.* Sw.

Bouteloua repens (H.B.K.) Scribn. & Merr.
    SYN = *B. filiformis* (Fourn.) Griffiths
Bromus diandrus Roth
    SYN = *B. rigidus* Roth
Bromus marginatus Nees
    SYN = *B. carinatus* H. & A.
Cenchrus ciliaris L.
    SYN = *Pennisetum ciliare* (L.) Link
Ceratoides lanata (Pursh) J. T. Howell
    SYN = *Eurotia l.* (Pursh) Moq.
Chasmanthium latifolium (Michx.) Yates
    SYN = *Uniola latifolia* Michx.
Chloris pluriflora (Fourn.) Clayton
    SYN = *Trichloris p.* Fourn.
Chrysothamnus viscidiflorus (Hook.) Nutt.
    SYN = *C. lanceolata* Nutt.
Cicuta douglasii (DC.) Coult. & Rose
    SYN = *C. occidentalis* Greene
Digitaria californica (Benth.) Henr.
    SYN = *Trichachne c.* (Benth.) Chase
Distichlis spicata (L.) Green
    SYN = *D. stricta* (Torr.) Rydb.
Gutierrezia sarothrae (Pursh) Britt. & Rusby
    SYN = *Xanthocephalum s.* (Pursh) Shimmers
Heterotheca villosa (Pursh) Shinners
    SYN = *Chrysopsis v.* (Pursh) Nutt. *ex* DC.
Koeleria pyramidata (Lam.) Beauv.
    SYN = *K. cristata* (L.) Pers.
              K. nitida Nutt.
Larrea tridentata (DC.) Cov.
    SYN = *L. divaricata* Cav.
Medicago hispida Gaertn.
    SYN = *M. polymorpha* L.
Plantago patagonica Jacq.
    SYN = *P. purshii* R. & S.
Poa fendleriana (Steud.) Vasey
    SYN = *P. longiligula* Scribn. & Williams
Poa sandbergii Vasey
    SYN = *P. secunda* Presl.
Ratibida columnaris (Sims) D. Don
    SYN = *R. columnifera* (Nutt.) Woot. & Standl.
Rhus trilobata Nutt.
    SYN = *R. aromatica* Ait.

Salsola iberica Sennen & Pau

    SYN = *S. kali* (of American authors)

Schizachyrium scoparium (Michx.) Nash

    SYN = *Andropogon scoparius* Michx.

            A. divergens Hitchc.

            A. littoralis Nash

Senecio longilobus Benth.

    SYN = *S. douglasii* DC.

Shepherdia canadensis (L.) Nutt.

    SYN = *Elaeagnus c.* (L.) A. Nels.

Sitanion hystrix (Nutt.) J. G. Smith

    SYN = *Elymus sitanion* Schult.

            E. elymoides Swezey

Sporobolus asper (Michx.) Kunth

    SYN = *S. macer* (Trin.) Hitchc.

Taeniatherum asperum (Simonkai) Nevski

    SYN = *Elymus caput-medusae* L. (as used in Hitchcock, 1951)

Vulpia octoflora (Walt.) Rydb.

    SYN = *Festuca o.* Walt.

Xanthocephalum dracunculoides (DC.) Shinners

    SYN = *Gutierrezia d.*

Selected References

Alley, Harold P. and Gary A. Lee. 1969. Weeds of Wyoming. Bull. 498. Agr. Exp. Sta., Univ. Wyoming, Laramie.

Agricultural Research Service. 1970. Selected weeds of the United States. Agr. Res. Service. United States Dep. Agr., Washington, D.C.

Andersen, Berniece A. Undated. Desert plants of Utah. Ext. Cir. 376. Coop. Ext. Service. Utah State Univ., Logan.

Andersen, Berniece A. and Arthur H. Holmgren. Undated. Mountain plants of northeastern Utah. Cir. 319. Coop. Ext. Service. Utah State Univ., Logan.

Babcock, Ernest Brown. 1947. The genus *Crepis*. Part two. Systematic treatment. Univ. California Press. Los Angeles.

Bare, Janet E. 1979. Wildflowers and weeds of Kansas. Regents Press of Kansas, Lawrence.

Barkley, T. M., Editor. 1977. Atlas of the flora of the Great Plains. Iowa State Univ. Press, Ames.

Beetle, Alan A. 1960. A study of sagebrush: The section *Tridentatae* of *Artemisia*. Bull. 368. Agr. Exp. Sta. Univ. Wyoming, Laramie.

Beetle, Alan A. 1977. Noteworthy grasses from Mexico. Phytologia 37:317– 407.

Beetle, Alan A. 1983. Las Gramineas de Mexico. Vol. I. COTECOCA.

Beetle, Alan A. and Morton May. 1971. Grasses of Wyoming. Res. J. 39. Agr. Exp. Sta., Univ. Wyoming, Laramie.

Benson, L. D. and R. A. Darrow. 1954. The trees and shrubs of southwestern deserts. Univ. New Mexico Press, Albuquerque.

Bentley, H. L. 1898. Grasses and forage plants of central Texas. Bull. 10. Div. Agrostology. United States Dep. Agr. Washington, D.C.

Best, Keith F., Jan Looman, and J. Baden Campbell. 1971. Prairie grasses. Pub. 1413. Canada Dep. Agr., Saskatchewan.

Booth, W. E. and J. C. Wright. 1959. Flora of Montana. Part II. Dicotyledons. Dept. of Botany and Microbiology, Montana State Univ., Bozeman.

Box, T. W., G. M. Van Dyne, and N. E. West. 1966. Syllabus on range resources of North America. Dep. Range Sci. Coll. Forestry and Renewable Resources. Colorado State Univ., Ft. Collins.

Budd, A. C. 1957. Wild plants of the Canadian Prairies. Pub. 983. Canada Dep. Agr., Saskatchewan.

Campbell, J. B., R. W. Lodge, and A. C. Budd. 1956. Poisonous plants of the Canadian Prairies. Pub. 900. Canada Dep. Agr., Ottawa.

Campbell, Robert S. and Wesley Keller. 1973. Range resources of the Southeastern United States. Amer. Soc. Agron. Spec. Pub. No. 21. Madison, Wisconsin.

Cassady, J. T. 1951. Bluestem range in the piney woods of Louisiana and east Texas. J. Range Manage. 4:173–177.

Clark, Lewis J. 1973. Wild flowers of British Columbia. Gray's Publ. Limited. Sidney, British Columbia.

Clausen, J., David D. Keck, and William Hiesey. 1940. Experimental studies on the nature of species. I. Effect of varied environments on western North American plants. (The *Achillea millefolium* complex.) Carnegie Inst. Pub. 520:296–324.

Coon, Nelson. 1979. Using plants for healing. Rodale Press, Emmaus, Pennsylvania.

Copple, R. F. and A. E. Aldous. 1932. The identification of certain native and naturalized grasses by their vegetative characters. Tech. Bull. 32. Agr. Exp. Sta., Kansas State Coll., Manhattan.

Copple, R. F. and C. P. Pase. 1967. A vegetative key to some common Arizona Range Grasses. Res. Paper RM-27. Rocky Mountain Forest and Range Exp. Sta. Forest Service. United States Dep. Agr., Washington, D.C.

Correll, Donovan S., and Marshall C. Johnson. 1970. Manual of the vascular plants of Texas. Texas Res. Found., Renner.

Cronin, Eugene H., and Darwin B. Nielson. 1979. The ecology and control of rangeland larkspurs. Bull. 499. Agr. Exp. Sta. Utah State Univ., Logan.

Cronquist, A. 1980. Vascular flora of the Southeastern United States. Vol. 1. Asteraceae. Univ. North Carolina Press. Chappel Hill.

Cronquist, A., A. H. Holmgren, N. H. Holmgren, J. R. Reveal, and P. K. Holmgren. 1977. Intermountain Flora. Vol. 6. Columbia Univ. Press, New York.

Cronquist, A., A. H. Holmgren, N. H. Holmgren, J. R. Reveal, and P. K. Holmgren. 1984. Intermountain Flora. Vol. 4. Columbia Univ. Press, New York.

Davis, Ray J. 1952. Flora of Idaho. William C. Brown Co., Dubuque, Iowa.

Dayton, William A. 1931. Important western browse plants. Misc. Pub. 101. United States Dep. Agr., Washington, D.C.

Densmore, Frances. 1974. How Indians use wild plants. Dover Publ., Inc., New York.

Dollahite, J. W., G. T. Housholder, and B. J. Camp. 1966. Oak poisoning in livestock. Bull. 1049. Agr. Exp. Sta. Texas A&M Univ., College Station.

Durrell, L. W. 1951. Halogeton—a new stock-poisoning weed. Circ. 170-A. Agr. Exp. Sta. Colorado State Coll., Ft. Collins.

Durrell, L. W., and I. E. Newsom. 1939. Colorado's poisonous and injurious plants. Bull. 455. Agr. Exp. Sta. Colorado State Coll., Ft. Collins.

Dyksterhuis, E. J. 1948. The vegetation of the Western Cross Timbers. Ecol. Monogr. 18:325–376.

Elias, Thomas S. 1980. Trees of North America. Van Nostrand Reinhold Co., New York.

Emboden, William. 1979. Narcotic plants. Macmillian Publ. Co., Inc., New York.

Epps, Alan C. 1976. Wild edible and poisonous plants of Alaska. Pub. 28. Coop. Ext. Service. Univ. Alaska, Fairbanks.

Evers, Robert A., and Roger P. Link. 1972. Poisonous plants of the Midwest and their effects on livestock. Spec. Pub. 24. Coll. Agr. Univ. Illinois, Urbana-Champaign.

Featherly, H. I. 1938. Grasses of Oklahoma. Tech. Bull. 3. Agr. Exp. Sta. Oklahoma Agr. and Mech. Coll., Stillwater.

Fernald, Merritt Lyndon. 1950. Gray's manual of botany. Amer. Book Co., New York.

Forest Service. 1937. Range plant handbook. Forest Service. United States Dep. Agr., Washington, D.C.

Frankton, Clarence. 1955. Weeds of Canada. Canada Dep. Agr., Ottawa.

Gates, Frank C. 1937. Grasses in Kansas. Vol. LV. No. 220-A. Kansas State Board of Agr., Topeka.

Gay, Charles W., Don Dwyer, Chris Allison, Stephan Hatch, and Jerry Schickedanz. 1980. New Mexico range plants. Circ. 374. Coop. Ext. Service. New Mexico State Univ., Las Cruces.

Gleason, Henry A. 1952. Illustrated flora of the northeastern United States and adjacent Canada. Lancaster Press, Inc., Lancaster, Pennsylvania.

Gould, Frank W. 1951. Grasses of southwestern United States. Biol. Sci. Bull. No. 7. Univ. Arizona, Tucson.

Gould, Frank W. 1975. Texas plants—a checklist and ecological summary. Misc. Pub. 585. Agr. Exp. Sta. Texas A&M Univ., College Station.

Gould, Frank W. 1975. The grasses of Texas. Texas A&M Univ. Press, College Station.

Gould, Frank W. 1978. Common Texas grasses. Texas A&M Univ. Press, College Station.

Gould, Frank W., and R. Moran. 1981. The grasses of Baja California, Mexico. Memoir 12. San Diego Soc. Natural History.

Gould, Frank W., and Robert B. Shaw. 1983. Grass systematics. 2nd Ed. Texas A&M Univ. Press, College Station.

Grinnell, George Bird. 1962. The Cheyenne Indians—their history and ways of life. Vol. 2. Cooper Square Publ. Inc., New York.

Hafenrichter, A. L., Lowell A. Mullen, and Robert L. Brown. 1949. Grasses and legumes for soil conservation in the Pacific Northwest. Misc. Pub. 678. United States Dep. Agr., Washington, D.C.

Harrington, H. D. 1954. Manual of the plants of Colorado. Sage Books, Denver.

Harrington, H. D. 1967. Edible native plants of the Rocky Mountains. Univ. New Mexico Press, Albuquerque.

Harrington, H. D. and L. W. Durrell. 1944. Key to some Colorado grasses in vegetative condition. Tech. Bull. 33. Agr. Exp. Sta. Colorado State Coll., Ft. Collins.

Harrington, H. D. and L. W. Durrell. 1957. How to identify plants. The Swallow Press Inc., Chicago.

Hart, Jeff and Jacqueline Moore. 1976. Montana-native plants and native people. Montana Hist. Soc., Helena.

Hayes, Doris W. and George A. Garrison. 1960. Key to important woody plants of eastern Oregon and Washington. Agr. Handbook 148. Forest Service. United States Dep. Agr., Washington, D.C.

Hermann, F. J. 1966. Notes on western range forbs. Agr. Handbook 293. Forest Service. United States Dep. Agr., Washington, D.C.

Hermann, F. J. 1970. Manual of the carices of the Rocky Mountains and Colorado Basin. Agr. Handbook 374. Forest Service. United States Dept. Agr., Washington, D.C.

Hilken, Thomas O., and Richard F. Miller. 1980. Medusahead (*Taeniatherum asperum Nevski*): A review and annotated bibliography. Sta. Bull. 644. Agr. Exp. Sta. Oregon State Univ., Corvallis.

Hitchcock, A. S. 1951. Manual of the grasses of the United States. Revised by Agnes Chase. Misc. Pub. 200. United States Dep. Agr., Washington, D.C.

Hitchcock, C. Leo. Undated. A key to the grasses of Montana. John S. Swift Co., Inc., St. Louis.

Holmgren, Arthur H. 1965. Handbook of vascular plants of the northern Wasatch. Nat. Press Co., Palo Alto, California.

Holmgren, Arthur H. and James L. Reveal. 1966. Checklist of the vascular plants of the Intermountain Region. Res. Paper INT-32. Forest Service. United States Dep. Agr. Washington, D.C.

Hosie, R. C. 1973. Native trees of Canada. Canadian Forestry Service. Dep. Environment. Ottawa.

Hulten, Eric. 1968. Flora of Alaska and neighboring territories. Stanford Univ. Press, Stanford, California.

Humphrey, Robert R. 1958. Arizona range grasses. Bull. 298. Agr. Exp. Sta. Univ. Arizona, Tucson.

Jepson, Willis L. 1951. Flowering plants of California. Univ. California Press, Berkeley.

Johnson, James R., and James T. Nichols. 1982. Plants of South Dakota grasslands. Bull. 566. Agr. Exp. Sta. South Dakota State Univ., Brookings.

Johnson, W. M. 1964. Field key to the sedges of Wyoming. Bull. 419. Agr. Exp. Sta. Univ. Wyoming, Laramie.

Jones, F. B. 1975. Flora of the Texas Coastal Bend. Mission Press, Corpus Christi, Texas.

Judd, B. Ira. 1962. Principal forage plants of Southwestern ranges. Sta. Paper 69. Rocky Mountain Forest and Range Exp. Sta. Forest Service. United States Dep. Agr. Washington, D.C.

Kartesz, John T., and Rosemarie Kartesz. 1980. A synonymized checklist of the vascular flora of the United States, Canada, and Greenland. Vol. II. The biota of North America. Univ. North Carolina Press, Chapel Hill.

Kearney, Thomas H. and Robert H. Peebles. 1960. Arizona flora. Univ. of California Press, Berkeley.

Keim, F. D., G. W. Beadle, and A. L. Frolik. 1932. The identification of the more important prairie grasses of Nebraska by their vegetative characteristics. Res. Bull. 65. Agr. Exp. Sta. Univ. Nebraska, Lincoln.

Kinch, Raymond C., Leon Wrage, and Raymond A. Moore. 1975. South Dakota weeds. Coop. Ext. Service, South Dakota State Univ., Brookings.

Kingsbury, John M. 1964. Poisonous plants of the United States and Canada. Prentice-Hall, Inc., Englewood Cliffs, New Jersey.

Knobel, Edward. 1980. Field guide to the grasses, sedges and rushes of the United States. Dover Publ. Inc., New York.

Kolstad, Ole A. 1974. Grasses of Nebraska. Memeo. Kearney State Coll., Kearney, Nebraska.

Langman, I. K. 1964. A selected guide to the literature on the flowering plants of Mexico. Univ. Pennsylvania Press, Philadelphia.

Leithead, Horace L., Lewis L. Yarlett, and Thomas N. Shiftlet. 1971. 100 native forage grasses in 11 southern states. Agr. Handbook 389. Soil Conserv. Service. United States Dep. Agr., Washington, D.C.

LeSueur, H. 1945. The ecology of the vegetation of Chihuahua, Mexico, north of parallel 28. Publ. 4521. Univ. Texas Press, Austin.

Lewis, Walter H. 1977. Medical botany. John Wiley & Sons, New York.

Little, Elbert L., Jr. 1971. Atlas of United States trees. Vol. 1. Conifers and important hardwoods. Misc. Publ. 1146. United States Dep. Agr., Washington, D.C.

Little, Elbert L., Jr. 1976. Atlas of United States trees. Vol. 3. Minor western hardwoods. Misc. Publ. 1314. United States Dep. Agr., Washington, D.C.

Lommasson, Robert C. 1973. Nebraska wild flowers. Univ. Nebraska Press, Lincoln.

Looman, J., and K. F. Best. 1979. Budd's flora of the Canadian prairie provinces. Pub. 1662. Res. Branch Agr. Canada.

Marsh, C. D. 1929. Stock-poisoning plants of the range. Bull. 1245. United States Dep. Agr., Washington, D.C.

Martin, S. Clark. 1975. Ecology and management of southwestern semi-desert grass-shrub ranges: The status of our knowledge. Res. Paper RM-156. Forest Service. United States Dep. Agr., Washington, D.C.

Martin, William C., and Charles R. Hutchins. 1980. A flora of New Mexico. Vol. 1. J. Cramer, Germany.

Martin, William C., and Charles R. Hutchins. 1981. A flora of New Mexico. Vol. 2. J. Cramer, Germany.

May, Morton. 1960. Key to the major grasses of the Big Horn Mountains—based on vegetative characters. Bull. 371. Agr. Exp. Sta. Univ. Wyoming, Laramie.

McKell, Cyrus M., James P. Blaisdell, and Joe R. Goodin. 1972. Wildland shrubs—their biology and utilization. Gen. Tech. Rep. INT-1. Forest Service. United States Dep. Agr., Washington, D.C.

Moreno, N. P. 1984. Glosario botanico ilustrado. Inst. Nac. de Invest. Xalapa, Veracruz, Mexico.

Morris, H. E., W. E. Booth, G. F. Payne, and R. E. Stitt. Undated. Important grasses on Montana ranges. Bull. 470. Agr. Exp. Sta. Montana State Coll., Bozeman.

Morton, Julia F. 1974. Folk remedies of the low country. E. A. Seemann Publ., Inc., Miami, Florida.

Muenscher, Walter Conrad. 1939. Poisonous plants of the United States. Macmillan Publ. Co., Inc., New York.

Munson, T. V. 1883. Forest and forest trees of Texas. Amer. J. Forestry. 1:433–451.

Munz, Philip A. and David D. Keck. 1959. A California flora. Univ. California Press, Berkeley.

Nebraska Statewide Arboretum. 1982. Common and scientific names of Nebraska plants. Publ. 101. Nebraska Statewide Arboretum, Lincoln.

Nelson, Ruth Ashton. 1968. Wild flowers of Wyoming. Bull. 490. Coop. Ext. Service. Univ. Wyoming, Laramie.

Owensby, Clinton E. 1980. Kansas prairie wildflowers. Iowa State Univ. Press, Ames.

Pammel, L. H., Carleton R. Ball, and F. Lamson-Scribner. 1904. The grasses of Iowa. Part II. Iowa Geol. Surv., Des Moines.

Parker, Karl G., Lamar R. Mason, and John F. Vallentine. Undated. Utah grasses. Ext. Circ. 384. Coop. Ext. Service. Utah State Univ., Logan.

Parks, H. B. 1937. Valuable plants native to Texas. Bull. 551. Agr. Exp. Sta. Texas A&M Univ., College Station.

Phillips, C. E. 1962. Some grasses of the Northeast. Field Manual 2. Agr. Exp. Sta. Univ. Delaware, Newark.

Phillips Petroleum Company. 1963. Pasture and range plants. Phillips Petroleum Co., Bartlesville, Oklahoma.

Pool, Raymond J. 1971. Handbook of Nebraska trees. Bull. 32. Conserv. Surv. Div. Univ. Nebraska, Lincoln.

Porter, C. L. 1960. Wyoming trees. Circ. 164R. Coop. Ext. Service. Univ. Wyoming, Laramie.

Preston, Richard J., Jr. 1976. North American Trees. Iowa State Univ. Press, Ames.

Rickett, Harold William. 1966. Wild Flowers of the United States. McGraw-Hill Book Co., New York.

Rydberg, P. Axel. 1932. Flora of the prairies and plains of central North America. Hafner Publ. Co., New York.

Rydberg, P. Axel. 1954. Flora of the Rocky Mountains and adjacent plains. Hafner Publ. Co., New York.

Rzedowski, J. 1978. Vegetacion de Mexico. Editorial Limusa, Mexico.

Sampson, Arthur W. 1924. Native American forage plants. John Wiley & Sons, New York.

Sampson, Arthur W., Agnes Chase, and Donald W. Hedrick. 1951. California grasslands and range forage grasses. Bull. 724. Agr. Exp. Sta. Univ. California, Berkeley.

Saunders, Charles Francis. 1976. Edible and useful wild plants. Dover Publ. Inc., New York.

Scoggan, H. J. 1979. The flora of Canada. Part 2. Pteridophyta, Gymnospermae and Monocotyledoneae. Museums of Canada, Ottawa.

Scoggan, H. J. 1979. The flora of Canada. Part 3. Dicotyledoneae (Saururaceae to Violaceae). Museums of Canada, Ottawa.

Scoggan, H. J. 1979. The flora of Canada. Part 4. Dicotyledoneae (Loasaceae to Compositae). Museums of Canada, Ottawa.

Shantz, H. L. and R. Zon. 1936. The natural vegetation of the United States. Pages 1–29. In Atlas Amer. Agr. Part 1, Section E. United States Dep. Agr., Washington, D.C.

Shaw, R. B., and J. D. Dodd. 1976. Vegetative key to the *Compositae* of the Rio Grande Plains of Texas. Misc. Pub. 1274. Agr. Exp. Sta. Texas A&M Univ., College Station.

Silveus, W. A. 1933. Texas grasses. Clegg Co., San Antonio.

Smeins, Fred E., and Robert B. Shaw. 1978. Natural vegetation of Texas and adjacent areas—1675–1975 bibliography. Misc. Pub. 1399. Agr. Exp. Sta. Texas A&M Univ., College Station.

Smith, James Payne, Jr. 1977. Vascular plant families. Mad River Press, Inc., Eureka, California.

Soil Conservation Service. 1965. Important native grasses for range conservation in Florida. Soil Conserv. Service, United States Dep. Agr., Washington, D.C.

Sperry, O. E., J. W. Dollahite, G. O. Hoffman, and B. J. Camp. 1977. Texas plants poisonous to livestock. Coop. Ext. Service. Texas A&M Univ., College Station.

State of Nebraska. 1962. Nebraska weeds. Bull. 101-R. Weed and Seed Div. Dept. of Agr. and Inspection, Lincoln.

Stechman, John V. 1977. Common western range plants. Vocational Educ. Productions. California Polytechnic State Univ., San Luis Obispo.

Stephens, H. A. 1973. Woody plants of the north central plains. Univ. Press of Kansas, Lawrence.

Stevens, O. A. 1963. Handbook of North Dakota plants. North Dakota Inst. Regional Studies. Fargo.

Steyermark, Julian A. 1963. Flora of Missouri. Iowa State Univ. Press, Ames.

Stoddart, L. A., A. H. Holmgren, and C. W. Cook. 1949. Important poisonous plants of Utah. Spec. Rep. No. 2. Agr. Exp. Sta. Utah State Univ., Logan.

Sutherland, David M. 1975. A vegetative key to Nebraska grasses. pp. 283–316. *In* prairie: "A multiple view. Univ. North Dakota Press, Grand Forks.

Texas Forest Service. 1963. Forest Trees of Texas. Texas Forest Service, College Station.

Tidestrom, I. 1925. Flora of Utah and Nevada. United States Nat. Herb. Contrib. 25:1–665.

Thilenius, John F. 1975. Alpine range management in the western United States—principles, practices, and problems: The status of our knowledge. Forest Service Res. Paper RM-157. United States Dep. Agr., Washington, D.C.

Tsvelev, N. N. 1984. Grasses of the Soviet Union. Russian Trans. Ser. 8. A. A. Balkema, Rotterdam.

Turner, B. L. 1959. The legumes of Texas. Univ. Texas Press, Austin.

Vallentine, John F. 1967. Nebraska range and pasture grasses. Ext. Circ. 76-170. Coop. Ext. Service. Univ. Nebraska, Lincoln.

Van Bruggen, Theodore. 1976. The vascular plants of South Dakota. Iowa State Univ. Press, Ames.

Villarreal-Q, J. A. 1983. Malezas de Buenavista Coahuila. Univ. Aut. Agraria "Antonio Narro." Buenavista, Saltillo, Mexico.

Vines, Robert A. 1960. Trees, shrubs and woody vines of the southwest. Univ. of Texas Press, Austin.

Wagner, Warren L. and Earl F. Aldon. 1978. Manual of the Saltbushes (*Atriplex* spp.) in New Mexico. Gen. Tech. Rep. RM-57. Forest Service. United States Dep. Agr., Washington, D.C.

Wasser, C. H. 1982. Ecology and culture of selected species useful in revegetating disturbed lands in the West. Fish and Wildlife Service, U.S. Dept. Interior, Washington, D.C.

Waterfall, U. T. 1962. Keys to the flora of Oklahoma. Oklahoma State Univ. Press, Stillwater.

Weiner, Michael A. 1980. Earth medicine-earth food. Macmillian Publ. Co., Inc., New York.

Western Regional Technical Committee W-90. 1972. Galleta: Taxonomy, ecology, and management of *Hilaria jamesii* on western rangelands. Bull. 487. Agr. Exp. Sta. Utah State Univ., Logan.

Williams, Kim. 1977. Eating wild plants. Mountain Press Publ. Co., Missoula, Montana.

Winward, A. H. 1980. Taxonomy and ecology of sagebrush in Oregon. Sta. Bull. 642. Agr. Exp. Sta. Oregon State Univ., Corvallis.

Appendix

Contest Rules
Society for Range Management
International Range Plant Identification Contest

ELIGIBILITY

Group 1. Teams: Each university may enter one team composed of three or four members. If the team consists of four members, the three highest scoring individuals will be considered the team for the university. The contestants must be enrolled undergraduates in any regular college course of study at the time of the contest or must have been enrolled as undergraduates in the fall semester or quarter immediately preceding the contest.

Group 2. Individuals: Any undergraduate student in any regular university course of study may compete for individual honors. Team members will be automatically entered in the individual competition.

SOURCE OF PLANT SPECIMENS

Study specimens A list of plants from the master plant list will be assigned to each participating college. Each university will be responsible for collecting and providing unmounted specimens of these plants to other participating universities. Collection data should be supplied with each specimen.

Contest specimens Plant specimens used for the contest will come from the SRM Range Plant Contest Herbarium. The chairman of the Plant Contest sub-committee of the Student Affairs Committee is responsible for the herbarium. Specimens for the herbarium are supplied by universities that participate in the contest. Each year the universities are requested to supply specimens of the plants on their assigned collection list. However, a school may supply specimens of any of the plants on the master plant list. Each specimen sent to the contest herbarium must be mounted on standard herbarium paper (11 x

16½ inches) with the name of the plant in the lower right hand corner. The identification of each specimen must be verified by a recognized taxonomist. The verification must be placed on each herbarium sheet. If possible, the stamp or seal of the authority or of the Herbarium employing the authority that verified the identification should be placed on each herbarium sheet.

CONTEST PROCEDURE

1. A minimum of one hundred (100) plants, appearing on the master list, will be selected by the Plant Contest subcommittee of the Student Affairs Committee.
2. Plant specimens will be mounted on standard 11 x 16 inch herbarium paper. The specimens will not be covered with any protective material.
3. Each specimen must show sufficient characteristics to identify it. Plant team coaches will view the contest specimens beginning approximately thirty (30) minutes prior to the contest. A specimen may be removed from the contest if one-third (⅓) of the coaches present vote in favor of removing the specimen.
4. Minimum distance between specimen sheets during the contest will be about twenty-two (22) inches.
5. Contestants will have fifty-five (55) seconds to write in the family or tribe and genus and species, and to check the longevity and origin columns. Five (5) seconds will be allowed to move to the next plant. Total elapsed time per plant will be one (1) minute.
6. Contestants will have three (3) minutes at the end of the contest to check their papers. Contestants will not be allowed to look at a plant a second time.
7. There will be no restriction on the number of duplicate mounts.
8. Contestants will not be permitted to handle the mounts. However, a hand lens may be used to assist in the identification of plants.
9. While the contest is in progress there shall be no conferring between contestants. Only the contestants and designated supervisors will be in the contest room during the contest.

CONTEST SCORING

1. The Plant Contest subcommittee will score the contestants papers. The papers will not be returned to the coaches or students.
2. Scoring will be as follows:
   a. Ten points are assigned to each plant as follows:

Family or tribe	2
Genus	3
Species	3
Longevity	1
Origin	1
Total	10

   b. Family or tribe only correct is 2 points.
   c. Genus only correct is 3 points.
   d. Family or tribe and genus correct is 5 points.
   e. Family or tribe, genus and species correct is 8 points.
   f. Family or tribe, genus, species and origin correct is 9 points.
   g. Longevity will be scored separately and is not tied to genus and species.
   h. The plant must be correctly identified as to genus to get credit for species.
   i. The plant must be correctly identified as to genus and species to get credit for origin.
   j. One point will be deducted for each family or tribe, genus or species that is correctly identified but misspelled.

CONTEST AWARDS

1. A rotating plaque will be awarded the first place team. This plaque becomes permanent property of the first school to win three times. The wins need not be in succession.
2. Team and individual awards will be given for the first five places. In the event of tie scores for team or individual places, duplicate awards will be given.
3. Plaques for permanent possession will be awarded to the first through fifth place teams.
4. The five highest scoring individuals will each be awarded a plaque for permanent possession.
5. Each contestant will receive a certificate.

CONTESTANT NO. _____

OFFICIAL SCORESHEET
Range Plant Identification Contest
Society for Range Management

PLANT NO.	ANNUAL	PERENNIAL	TRIBE OR FAMILY	GENUS	SPECIFIC EPITHET	NATIVE	INTRODUCED

Index

462